NASA STI Program . . . in Profile

Since its founding, NASA has been dedicated to the advancement of aeronautics and space science. The NASA Scientific and Technical Information (STI) program plays a key part in helping NASA maintain this important role.

The NASA STI Program operates under the auspices of the Agency Chief Information Officer. It collects, organizes, provides for archiving, and disseminates NASA's STI. The NASA STI program provides access to the NASA Aeronautics and Space Database and its public interface, the NASA Technical Reports Server, thus providing one of the largest collections of aeronautical and space science STI in the world. Results are published in both non-NASA channels and by NASA in the NASA STI Report Series, which includes the following report types:

- TECHNICAL PUBLICATION. Reports of completed research or a major significant phase of research that present the results of NASA programs and include extensive data or theoretical analysis. Includes compilations of significant scientific and technical data and information deemed to be of continuing reference value. NASA counterpart of peer-reviewed formal professional papers but has less stringent limitations on manuscript length and extent of graphic presentations.

- TECHNICAL MEMORANDUM. Scientific and technical findings that are preliminary or of specialized interest, e.g., quick release reports, working papers, and bibliographies that contain minimal annotation. Does not contain extensive analysis.

- CONTRACTOR REPORT. Scientific and technical findings by NASA-sponsored contractors and grantees.

- CONFERENCE PUBLICATION. Collected papers from scientific and technical conferences, symposia, seminars, or other meetings sponsored or cosponsored by NASA.

- SPECIAL PUBLICATION. Scientific, technical, or historical information from NASA programs, projects, and missions, often concerned with subjects having substantial public interest.

- TECHNICAL TRANSLATION. English-language translations of foreign scientific and technical material pertinent to NASA's mission.

Specialized services also include creating custom thesauri, building customized databases, organizing and publishing research results.

For more information about the NASA STI program, see the following:

- Access the NASA STI program home page at *http://www.sti.nasa.gov*

- E-mail your question via the Internet to *help@sti.nasa.gov*

- Fax your question to the NASA STI Help Desk at 443–757–5803

- Telephone the NASA STI Help Desk at 443–757–5802

- Write to:
  NASA Center for AeroSpace Information (CASI)
  7115 Standard Drive
  Hanover, MD 21076–1320

NASA/TM—2010-216764

Interplanetary Mission Design Handbook:
Earth-to-Mars Mission Opportunities 2026 to 2045

Laura M. Burke, Robert D. Falck, and Melissa L. McGuire
Glenn Research Center, Cleveland, Ohio

National Aeronautics and
Space Administration

Glenn Research Center
Cleveland, Ohio 44135

October 2010

This report is a formal draft or working paper, intended to solicit comments and ideas from a technical peer group.

Level of Review: This material has been technically reviewed by technical management.

Available from

NASA Center for Aerospace Information
7115 Standard Drive
Hanover, MD 21076–1320

National Technical Information Service
5301 Shawnee Road
Alexandria, VA 22312

Contents

Introduction	1
Nomenclature	1
Trajectory Characteristics	2
Ballistic Trajectories	4
Transfer Trajectories Using Deep Space Maneuvers	7
Launch/Injection Geometry	8
Launch Azimuth	8
Description of Trajectory Characteristics	8
Mission Opportunities	9
Assumptions	11
Mission Design Data Contour Plots	12
Earth to Mars Mission Opportunities 2026 to 2045	12
Earth to Mars—2026 Opportunity	12
Earth to Mars—2028 Opportunity	19
Earth to Mars—2031 Opportunity	26
Earth to Mars—2033 Opportunity	33
Earth to Mars—2035 Opportunity	40
Earth to Mars—2037 Opportunity	47
Earth to Mars—2039 Opportunity	54
Earth to Mars—2041 Opportunity	61
Earth to Mars—2043 Opportunity	68
Earth to Mars—2045 Opportunity	75
Appendix A.—Verification of Midas Results	82
Verification Contour Plots for 2005 Opportunity	83
References	85

List of Tables

Table 1.—Data for optimal missions: 2026 to 2045	10
Table 2.—Earth to Mars—2026 Opportunity—Energy Minima	12
Table 3.—Earth to Mars—2028 Opportunity—Energy Minima	19
Table 4.—Earth to Mars—2031 Opportunity—Energy Minima	26
Table 5.—Earth to Mars—2033 Opportunity—Energy Minima	33
Table 6.—Earth to Mars—2035 Opportunity—Energy Minima	40
Table 7.—Earth to Mars—2037 Opportunity—Energy Minima	47
Table 8.—Earth to Mars—2039 Opportunity—Energy Minima	54
Table 9.—Earth to Mars—2041 Opportunity—Energy Minima	61
Table 10.—Earth to Mars—2043 Opportunity—Energy Minima	68
Table 11.—Earth to Mars—2045 Opportunity—Energy Minima	75
Table 12.—Energy minima for 2005 opportunity calculated by MIDAS	82
Table 13.—Energy minima for 2005 opportunity data from Reference 6	82

List of Figures

Figure 1.—Minimum departure energies for Earth to Mars Ballistic Missions: 1990 to 2045.	3
Figure 2.—Lambert Theorem Geometry.	4
Figure 3.—Mission space in departure/arrival date coordinates.	5
Figure 4.—Target planet orbital plane geometry forcing a polar inclination for a 180° transfer.	6
Figure 5.—Mission space with nodal transfer.	6

Figure 6.—Orbit geometry for nodal transfer. 7
Figure 7.—Type I optimal ballistic mission departure energies. 9
Figure 8.—Type II optimal ballistic mission departure energies. 10

Interplanetary Mission Design Handbook:
Earth-to-Mars Mission Opportunities 2026 to 2045

Laura M. Burke, Robert D. Falck, and Melissa L. McGuire
National Aeronautics and Space Administration
Glenn Research Center
Cleveland, Ohio 44135

Introduction

The purpose of this Mission Design Handbook is to provide trajectory designers and mission planners with graphical information about Earth to Mars trajectory opportunities for the years of 2026 through 2045. The trajectory data used to create the following opportunity contour plots was generated using MIDAS, a patched conic interplanetary trajectory optimization program that is able to optimize the times of specified trajectory events and other trajectory parameters (Ref. 1). The contour plots themselves were generated using the data visualization capabilities of MATLAB (The Mathworks, Inc.). The plots, displayed on a departure date/arrival date mission space, show departure energy, right ascension and declination of the launch asymptote, and target planet hyperbolic arrival excess speed, V_∞, for each launch opportunity.

Trajectory contour plots are particularly important in the beginning stages of mission design as valuable tools that display the interplanetary flight path characteristics for a particular launch opportunity to Mars. The use of these contour plots is an important first step for determining initial optimal launch opportunities for interplanetary missions. They also serve as good approximations for directional values of the launch asymptote vector, target planet (Mars) arrival excess velocities, and total mission flight time. These plots allow a mission designer to determine the basic requirements for an Earth to Mars transfer vehicle as well as a preliminary estimate of the required propellant load.

Provided in this study are two sets of contour plots for each launch opportunity. The first set of plots shows Earth to Mars ballistic trajectories without the addition of any deep space maneuvers. The second set of plots shows Earth to Mars transfer trajectories with the addition of deep space maneuvers, which further optimize the determined trajectories. Providing two sets of plots for each opportunity allows mission planners the ability to compare and contrast different mission architectures.

Nomenclature

C_3	Earth departure energy (km^2/sec^2), equal to the square of the departure hyperbolic excess velocity
d_{J2000}	launch date in terms of full integer days elapsed since Jan. 1, 2000 $1^d\ 12^h$ UT (JD = 2451545)
$DVMT$	total magnitude of the sum of the deep space maneuvers (km/sec)
GHA_{DATE}	Greenwich hour angle at 0 h GMT of any date, assumes equator is J2000 (deg)
h	altitude (km)
TOF	time of flight (days)
t_L	time of launch (h, GMT, i.e., mean solar time)
V	spacecraft velocity of Mars flyby (km/sec)
V_∞	hyperbolic excess velocity (km/sec)

α_∞	right ascension of the launch asymptote (deg)
α_L	right ascension of the launch site (deg)
ΔV	delta velocity (km/sec)
δ_∞	declination of the launch asymptote (deg)
γ	Vernal Equinox
λ_L	east longitude of launch site (deg)
μ_M	GM, gravitational parameter of Mars, 42,828.3 km³/sec²
ϕ_L	geocentric latitude of the launch site (deg)
r_M	radius of Mars, 3,397 km
Σ_∞	launch azimuth (deg)
ω_{EARTH}	inertial rotation rate of Earth, 15.041067179 deg/h of mean solar time
ω_E	angular velocity of Earth, 1.994×10^{-7} rad/s
ω_M	angular velocity of Mars, 1.06×10^{-7} rad/s
τ_S	synodic period (days)

Trajectory Characteristics

For the purpose of minimizing the required transfer energy, Earth departure and Mars arrival should occur when the two planets are in conjunction. Conjunction class missions, or Hohmann transfer missions, occur when the Earth at launch and Mars at arrival are essentially on opposite sides of the sun.

The trajectories calculated with MIDAS were specified to be Venus flyby trajectories. A flyby arrival event was chosen because no weight is placed on the arrival ΔV thereby reducing the trajectory's total ΔV. A propulsive capture arrival event was not chosen because it requires a large ΔV maneuver near the arrival planet and thereby obscures the Earth relative performance requirements.

Using the hyperbolic excess speed, V_∞, the gravitational parameter μ_M, and radius r_M, and the spacecraft's altitude h of flyby, the velocity of a Mars flyby trajectory is given by:

$$V = \sqrt{\frac{2\mu_M}{r_M + h} + V_\infty^2} \tag{1}$$

Optimal launch dates for minimum departure energy conjunction class missions reoccur every synodic period. The synodic period is the time required for any phase angle to repeat, which for Mars with respect to the Earth is 779.935 Earth days (approximately 2.14 years), and can be confirmed using the following equation (Ref. 2):

$$\tau_S = \frac{2\pi}{|\omega_E - \omega_M|} \tag{2}$$

where ω_E and ω_M are the rotational rates of the Earth and Mars about the Sun.

Due to the fact that Earth and Mars orbits are neither exactly circular nor coplanar, the trajectory characteristics of each opportunity are not always the same. One opportunity may require less departure energy and have a lower Mars arrival V_∞ than another opportunity. However, Earth and Mars nearly return to their original relative heliocentric positions every 7 to 8 synodic periods, or every 15 to 17 years (Ref. 3). Figure 1 shows the repeating cycle of minimum (either Type I or Type II) departure energy values for Earth to Mars ballistic opportunities 1990 to 2045. Type I trajectories are characterized as having shorter trip times and Type II trajectories are characterized as having longer trip times usually with a lower required ΔV than Type I trajectories. Type I and Type II trajectories have heliocentric travel angles less than and greater than 180° respectively and are discussed in more detail in the Ballistic Trajectories section of the handbook.

Each pair of departure/arrival dates defines a unique Earth to Mars transfer trajectory. For the purposes of these particular contour plots, each date pair is associated with an array of specific values for departure energy, right ascension and declination of the launch asymptote, and Mars arrival V_∞. The resulting contours for each specified parameter are plotted in an Earth departure/Mars arrival mission space with a departure date coverage span of 160 days and an arrival date coverage span of 400 days. Since numerous events can cause a mission to launch at a date other than optimal, the plots include departure energies up to 50 km^2/sec^2. Departure energies above 50 km^2/sec^2 were considered to be generally not of interest because of the large propulsive maneuvers they require. A 160-day span for departure date was used because it sufficiently covered the range of desired departure energies. The 400-day coverage span for Mars arrival dates was used in order to display time-of-flight times ranging from 100 to 450 days.

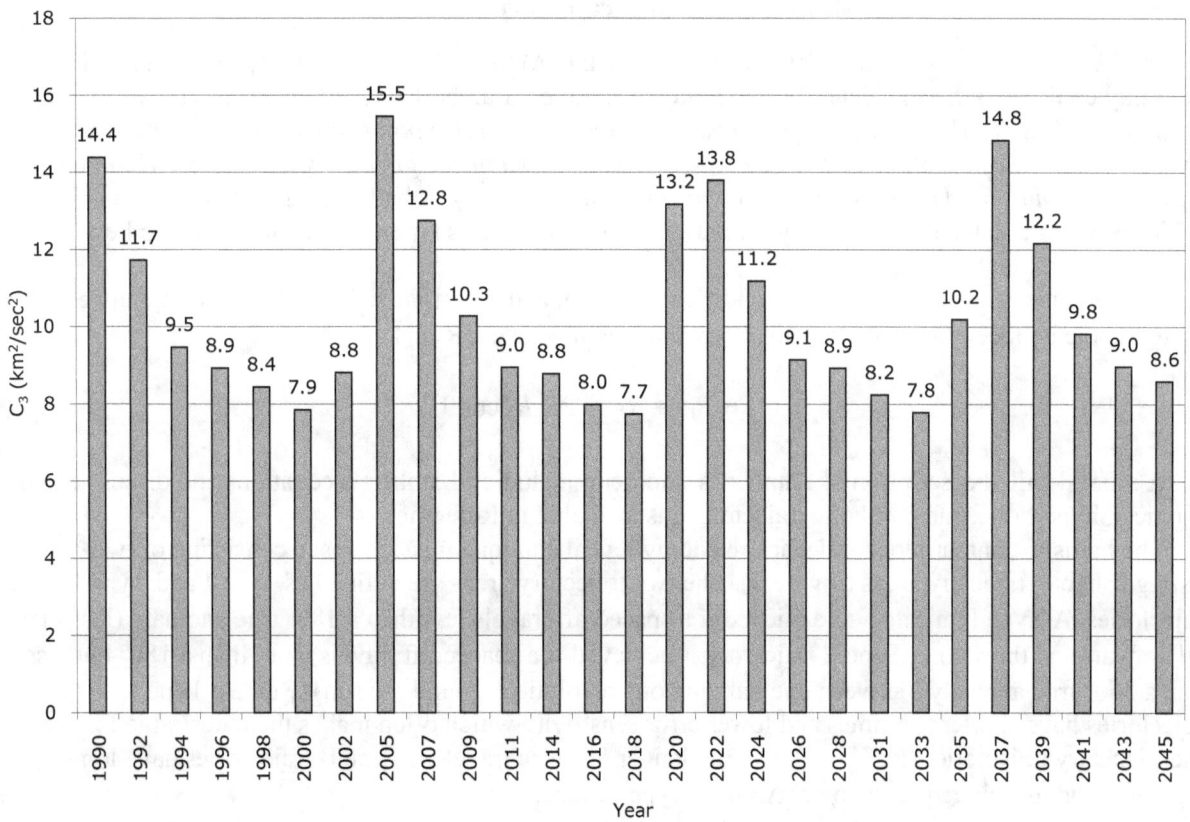

Figure 1.—Minimum departure energies for Earth to Mars Ballistic Missions: 1990 to 2045.

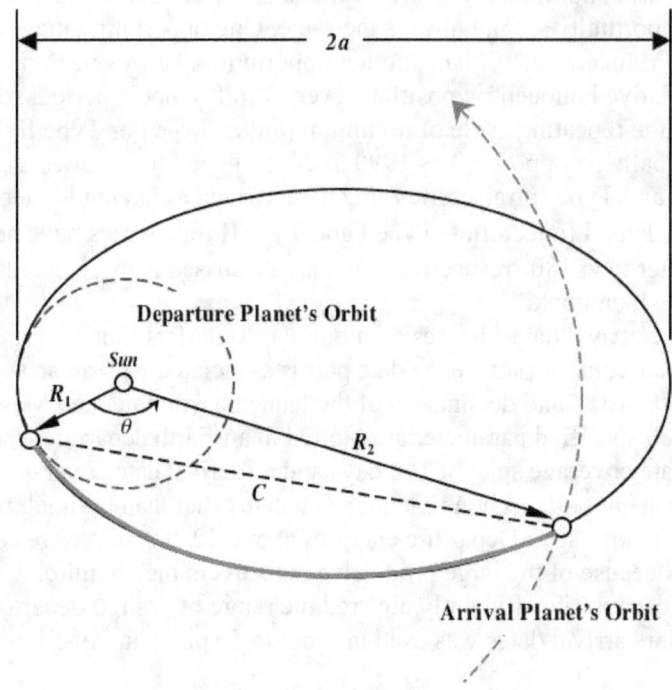

Figure 2.—Lambert Theorem Geometry.

Ballistic Trajectories

MIDAS uses Lambert's Theorem to calculate Earth to Mars ballistic transfer trajectories in which two-body conic motion and a central force field are assumed. Lambert's Theorem defines the following relationship (Ref. 4): *the transfer time of a body moving between two points on a conic trajectory is a function only of the sum of the distances of the two points from the origin of force, the linear distance between the points, and the semi-major axis of the conic* (Figure 2). A ballistic mission to Mars can be sufficiently represented by a two-body formation, thus, making this type of mission suitable to be analyzed using Lambert's method.

In most interplanetary missions, R_1 and R_2 are known, and the distance between them, C, can be related to the heliocentric transfer angle, θ, by the law of cosines (Ref. 4):

$$C^2 = R_1^2 + R_2^2 - 2R_1R_2 \cos\theta \tag{3}$$

A more detailed description of Lambert's Theorem, including Lambert's equations and discussions of elliptical, hyperbolic, and parabolic trajectories, is available in Reference 4.

The ballistic contour plots of departure energy reveal a unique mission space consisting of two distinguishable trajectory areas (Figure 3). The two trajectory areas are defined as Type I and Type II trajectories. A Type I trajectory is achieved if a spacecraft travels less than a 180° true anomaly (less than halfway around the sun); a Type II trajectory is achieved if a spacecraft travels more than a 180° but less than a 360° true anomaly (between one half and one revolution around the sun) (Ref. 3). Type I trajectories have shorter trip times and lower error sensitivity which often makes them desirable for interplanetary trajectories (Ref. 5). Due to the longer distance traveled, Type II trajectories have longer trip times and usually require a lower ΔV than Type I trajectories.

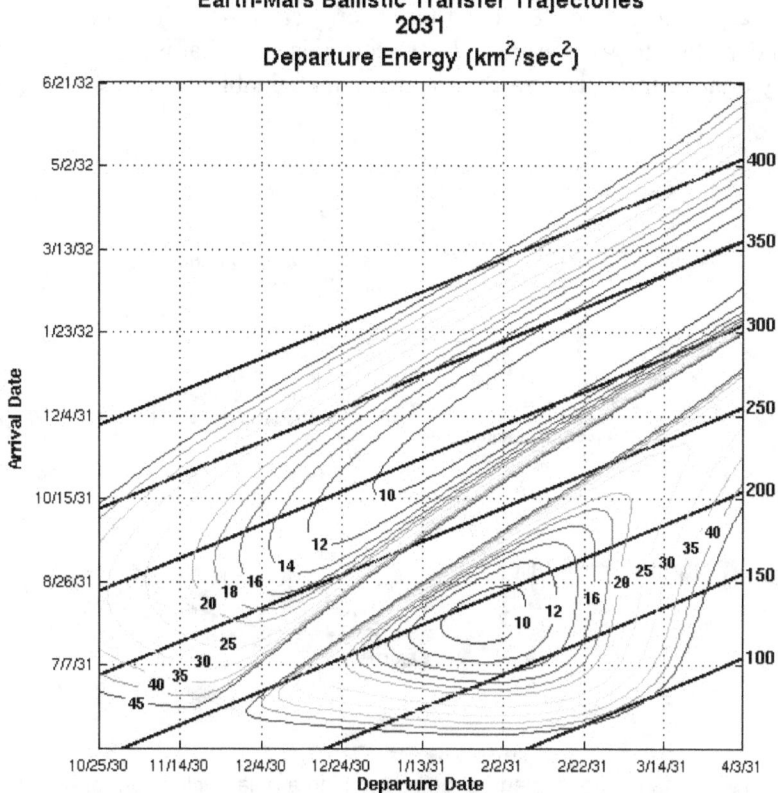

Figure 3.—Mission space in departure/arrival date coordinates.

It should be noted that there are two minimum energy areas within the ballistic plots, one of which is associated with a Type I transfer and the other with a Type II transfer (Ref. 3). The values of the energy minima for each opportunity are summarized in the Mission Opportunities section of this handbook.

A notable ridge passes diagonally from the lower left to the upper right of the mission space separating the Type I and Type II trajectories. This dramatic rise is attributed to near-180° transfer angle trajectories. The cause of this phenomenon is that, for a ballistic 180° trajectory, the Sun and both trajectory end points must lie in the plane of the transfer orbit, and the endpoints must lie in the orbital plane of the departure planet along a common diameter in the ecliptic. If the nodal lines of the target planet and the transfer orbit are not aligned, any slight inclination by the target planet to cause a vertical out of plane displacement forces a polar inclination for a 180° transfer in order to recover (Ref. 6). The presence of an arrival position error (Figure 4), caused by inclination of the target planet's orbit, at the Mars arrival location opposite the Earth departure location signifies that a polar trajectory must be used in order to reach the target planet at that point. Increasing the inclination, i, of the trajectory plane to anything less than 90° will not reach the target planet. Figure 4 describes the geometry of the transfer orbit, arrival planet orbit, and departure planet orbit that forces a polar 180° transfer. The dotted red lines show the transfer orbit's progression as the inclination of the orbit is increased to 90° in order to arrive at to the target planet in this current geometry. Near-180° transfer trajectories require larger departure energies than orbits with less inclination because they are not able to take advantage of the energy provided by Earth's orbital velocity. Hence, the Earth to Mars Transfer vehicle becomes responsible for obtaining the necessary orbital velocity.

An exception to the polar inclination 180° transfer trajectories that occur in all ballistic missions is a nodal transfer (Figure 5). Nodal transfer opportunities are identified by a single node (highlighted in red in Figure 5) which connects the Type I and Type II trajectory peaks. A nodal transfer occurs when Earth departure takes place at the node where the orbit of the target planet (Mars) intersects the Earth's orbit, and arrival occurs at the opposite such node (Figure 6). In these cases, the transfer trajectory plane can lie in Earth's orbital plane, which requires less departure energy (Ref. 6). While the trajectory is still swept out 180° from Earth departure to Mars arrival, the trajectory is such that it provides an opportunity for a low departure energy transfer instead of trajectories with unattainable departure energy levels.

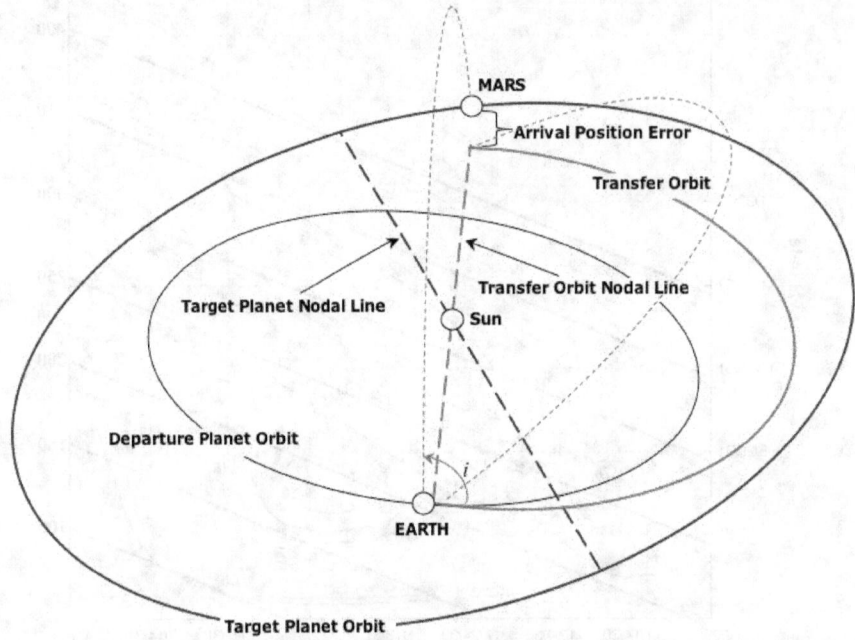

Figure 4.—Target planet orbital plane geometry forcing a polar inclination for a 180° transfer.

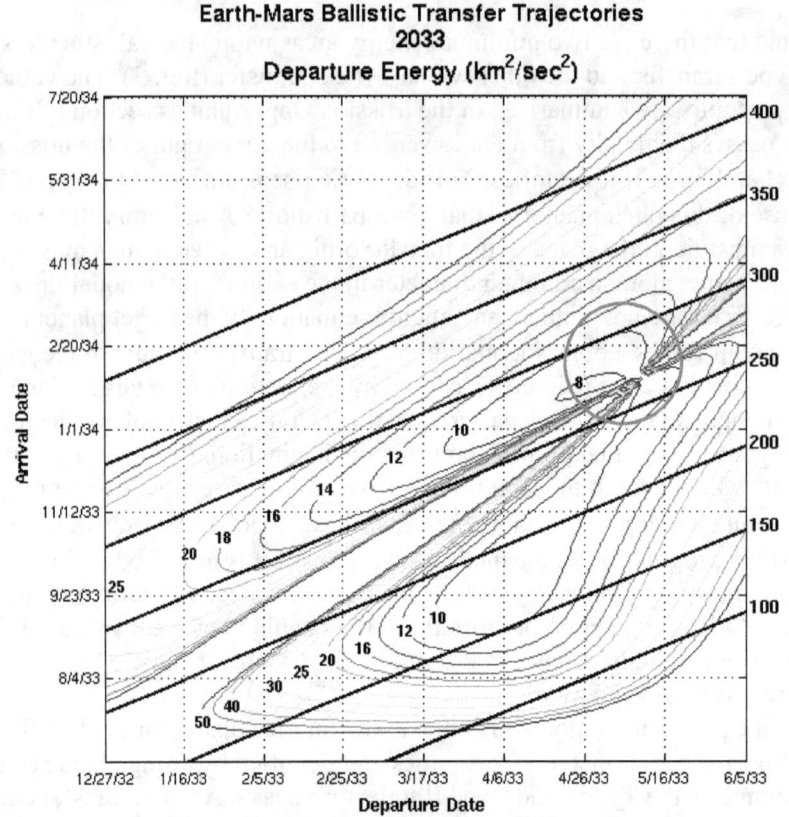

Figure 5.—Mission space with nodal transfer.

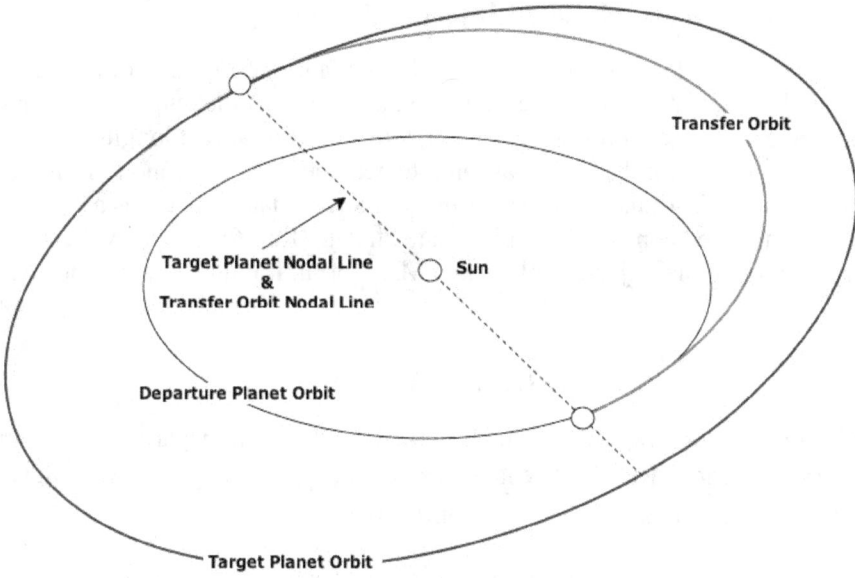

Figure 6.—Orbit geometry for nodal transfer.

While nodal transfers present a low departure energy advantage, they represent single-time-point missions with extraordinarily high error sensitivity, which is generally undesirable in current missions. A nodal transfer opportunity is not present in all opportunity cases, as demonstrated by the large ridge in Figure 3.

Transfer Trajectories Using Deep Space Maneuvers

The addition of deep space maneuvers changes the architecture of a trajectory and may well reduce the overall cost of the mission by potentially minimizing the necessary total initial mass of a launch vehicle or the total ΔV of a mission (Ref. 7). Departure energies and arrival excess speeds are significantly reduced along the ridge in Figure 3 with the addition of deep space maneuvers.

MIDAS determines where to perform a deep space maneuver by using calculus of variations and primer vector theory. If MIDAS determines that a deep space maneuver reduced ΔV, it places that maneuver at or near a node. This burn is often called a "broken plane" maneuver. The goal of performing broken plane maneuvers is to avoid high ecliptic inclination of the trajectory by performing a plane change maneuver, such that it would correct the path of the spacecraft toward the target planet's out-of-ecliptic position. The affects of performing this type of maneuver can be observed in the plots of departure energy for missions with added deep space maneuvers by observing the lack of a ridge of high departure energies separating Type I and Type II trajectories.

Optimal trajectories for the launch opportunities with deep space maneuvers are not listed because in order to determine the minima ΔV case, the Earth departure ΔV must be combined with the total ΔV of the deep space maneuvers. A contour plot of Earth departure ΔV is not provided because Earth departure ΔV is dependent on initial departure orbit. However, Earth departure ΔV can be computed from the plot of departure energy and then combined with the magnitudes of the deep space maneuvers to reveal the minimum ΔV trajectory.

The contour plots of trajectories that utilize deep space maneuvers are of importance for determining Earth to Mars transfer vehicle requirements and launch asymptote specifications as well as verifying that a particularly low value for the total ΔV does not correspond to an unusually high departure energy.

Launch/Injection Geometry

The launch V_∞ vector is referenced to an Earth Ecliptic plane and Equinox of the year 2000 coordinate system. The declination, δ_∞, corresponds to the latitude of the outgoing asymptote from the equator. The right ascension, α_∞, represents the asymptote's equatorial east longitude from the vernal equinox. A graphical depiction of the launch asymptote geometry can be found in Reference 6.

The outgoing V_∞ vector is a function of departure and arrival date. Since it is a slowly varying function, it may be considered constant for a given launch date (Ref. 6). The arrival excess speed vector is located at the intersection of the trajectory plane and Mars' orbit. It points outward normal to the Earth's surface.

Launch Azimuth

An optimal Earth departure trajectory plane is defined by the Earth departure V_∞ vector, the launch site, and the center of the Earth (Ref. 6). The launch azimuth, Σ_L, the angle between the trajectory plane and the launch meridian, specifies the orientation of this plane:

$$\cotan(\Sigma_L) = \frac{\cos(\phi_L)\tan(\delta_\infty) - \sin(\phi_L)\cos(\alpha_\infty - \alpha_L)}{\sin(\alpha_\infty - \alpha_L)} \tag{3}$$

The launch azimuth can also be defined in terms of the arc between the launch site and the arrival excess speed vector (Ref. 6), θ:

$$\cos(\Sigma_L) = \frac{\sin(\delta_\infty) - \sin(\phi_L)\cos(\theta)}{\cos(\phi_L)\sin(\theta)} \tag{4}$$

For a launch site of the NASA Kennedy Space Flight Center at Cape Canaveral, Florida, which has a declination of 28.45°, the Eastern Test Range Safety Requirements specify that the launch azimuth must be between 40° and 115°. This condition affects the times at which spacecrafts can be launched from this particular position (Ref. 2).

Right ascension of the launch sire and Greenwich hour angle are two other quantities that define the trajectory plane and can be obtained from the following approximate expressions (Ref. 6):

$$\alpha_L = \lambda_L + GHA_{DATE} + \omega_{EARTH} \cdot t_L \tag{5}$$

$$GHA_{DATE} = 100.4606 + 0.985647365 \cdot d_{J2000} \tag{6}$$

where
$\quad d_{J2000}$ is the number of days past noon January 1, 2000, UTC
$\quad \omega_{EARTH}$ is the Earth's rotation rate about its axis
$\quad \lambda_L$ is the launch site longitude

For information concerning daily launch windows consult Reference 6.

Description of Trajectory Characteristics

The plots are presented in a departure date/arrival date mission space. For each opportunity four parameters, including departure energy, launch asymptote declination and right ascension, and Mars arrival V_∞ are each plotted within the mission space. These parameters are presented in two sets of contour plots, one without a performed deep space maneuver to minimize the departure energy, the ballistic trajectories, and another with a performed deep space maneuver.

For the deep space maneuver option missions, a contour plot of the total magnitude of the deep space maneuvers as well as contour plots of the time and magnitude of each individual deep space maneuver is included in addition to the contour plots of the four parameters plotted for the ballistic missions. Included values of the total magnitude range from 0.0 to 5 km/sec. A value of zero signifies that no deep space maneuvers were performed. As such, some trajectories that lie outside of the outermost contour (0.0 km/sec) are not subject to any deep space maneuvers.

Only the contour plots of the departure energy include diagonal time-of-flight lines. The time-of-flight lines occur in 50-day increments. The contour plots of right ascension and declination of the launch asymptote as well as the plots of Mars arrival V_∞ are superimposed on plots of the corresponding departure energy for reference purposes. The contour lines of the parameters are labeled with their values.

A table summarizing the optimal energy trajectories for both Type I and Type II ballistic missions is given at the beginning of each opportunity's set of contour plots. This summary table provides the optimal mission characteristic values for both minimum departure energy and minimum Mars arrival V_∞. Each row in the table defines a particular optimized mission. The trajectory characteristic optimized for that mission is in bold type.

Mission Opportunities

The departure energy for the optimal mission was calculated for each launch opportunity for both Type I and Type II ballistic transfer trajectories. Summaries of these values are provided in Figure 7 and Figure 8.

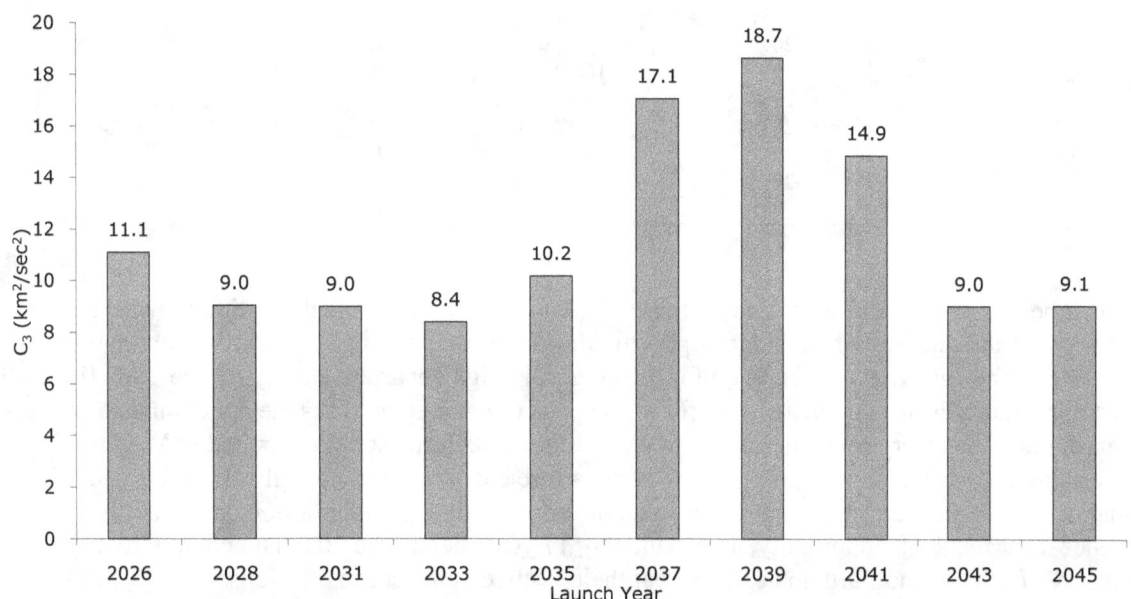

Figure 7.—Type I optimal ballistic mission departure energies.

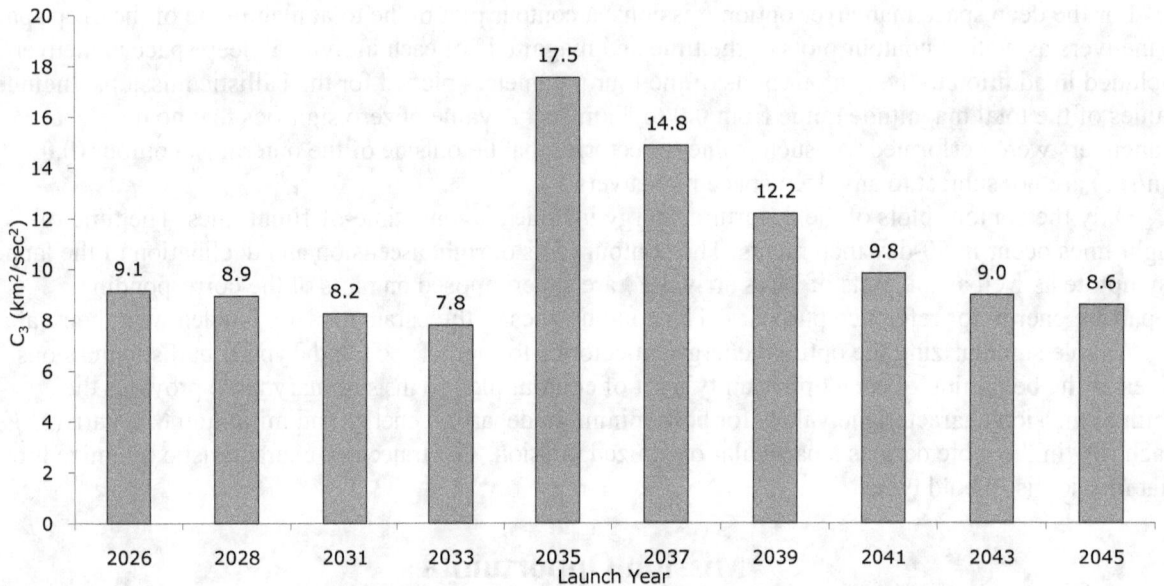

Figure 8.—Type II optimal ballistic mission departure energies.

TABLE 1.—DATA FOR OPTIMAL MISSIONS: 2026 TO 2045

Year	C_3 (km^2/sec^2)	Transfer type
2026	9.144	II
2028	8.928	II
2031	8.237	II
2033	7.781	II
2035	10.19	I
2037	14.84	II
2039	12.17	II
2041	9.818	II
2043	8.969	II
2045	8.587	II

It can be observed from Figure 7 and Figure 8 that a Type II mission is the optimal lowest energy trajectories for the majority of the launch opportunities. However, a Type I trajectory is optimal for the 2035 launch opportunity. An explanation for the unusually high departure energy for the Type II transfer for this opportunity is that the arrival date for the Type II transfer approaches the aphelion date of Mars, and larger departure energies are necessary to reach Mars at its furthest point. Arrival at Mars before this date by using a Type I trajectory provides lower departure energies, consequently, Type I trajectories are optimal in these instances. This event can also be observed in the optimal mission data provided by Reference 3 for the 2018 opportunity, approximately 17 years before the 2035 launch opportunity, the time it takes for Mars and Earth to nearly repeat their relative heliocentric positions.

As previously stated, some year's optimal launch opportunity may be more optimal than another year's optimal launch opportunity because the orbits of Earth and Mars are neither exactly circular nor coplanar. For the launch opportunities in this study, the opportunity requiring the least amount of departure energy occurs at 2033.

Assumptions

The optimization code in MIDAS is defaulted to minimize total weighted mission ΔV therefore only conjunction class missions were considered in this study.

Departure energy, declination and right ascension of the launch asymptote, and Mars arrival V_∞ are assumed to be the same for flyby and capture trajectories when no deep space maneuvers are introduced into the mission. The rationalization for this is that the same arrival position is always targeted in a ballistic trajectory because there is no additional directional control over the spacecraft once it departs Earth. Hence, the V_∞ vector will be the same regardless of the arrival event, and consequently the right ascension and declination of the V_∞ vector remain the same. For missions in which deep space maneuvers are included, specifying a flyby trajectory over a capture trajectory does have consequences on the mission characteristics. The values of departure energy, declination and right ascension of the departure asymptote, and Mars arrival V_∞ are all affected by changing the arrival event. The differences in the trajectory parameters caused by specifying either a flyby or capture arrival event are the result of optimizing the deep space maneuver to reduce the overall mission ΔV. In trying to minimize Mars arrival ΔV for a capture arrival event, MIDAS calculates the optimal trajectory with a deep space maneuver in order to reach Mars with the optimal conditions for capture. Since Mars arrival ΔV is now being accounted for by the deep space maneuver, the original outgoing V_∞ direction is changed to place the deep space maneuver in the optimal location at the optimal time, which may be drastically different than the placement of the outgoing V_∞ vector when a flyby arrival event is specified.

The Earth to Mars flight paths without added deep space maneuvers are ballistic trajectories, meaning, after the initial injection burn the spacecraft coasts the remainder of the trip to Mars. The Earth to Mars flight paths with added deep space maneuvers are, however, not ballistic. These trajectories rely on mid-course impulsive burns to optimize various trajectory characteristics en route to Mars. For missions in which maneuvers were added, no more than two were actually performed. MIDAS is capable of adding up to eight deep space maneuvers into a trajectory, however, using just two usually accomplishes the desired departure energy reduction without over complicating mission trajectories.

For all contour plots, the spacecraft departs from a 407 km circular parking orbit. Inclination was not independently set for this study, as MIDAS has a parking orbit inclination default value of 90°. The default value of 90° was used because there is no plane change required, and therefore no ΔV required, in order to depart from a lesser inclination.

In order to generate a range of data for each parameter, a parameter search was performed in MIDAS in which the departure and arrival dates were varied over a set range in two-day increments. MIDAS calculates the values of the specified trajectory characteristics only to a precision of two decimal places. In some instances, this complicates the task of determining exactly which date is optimal for launch because two or more dates may result in having the same departure energy due to MIDAS rounding.

Mission Design Data Contour Plots

Earth to Mars Mission Opportunities 0 to 20

Earth to Mars—2026 Opportunity

TABLE 2.—EARTH TO MARS—2026 OPPORTUNITY—ENERGY MINIMA

Mission type	Earth departure date (m/d/yr)	Mars arrival date (m/d/yr)	C_3 (km²/sec²)	Right ascension (deg)	Declination (deg)	Mars arrival excess speed (km/s)
Type 1	11/14/26	8/9/27	11.11	120	28.28	2.915
Type 2	10/31/26	8/19/27	9.144	130.7	23.16	2.729
Type 1	11/14/26	8/9/27	11.11	120	28.28	2.915
Type 2	11/6/26	9/8/27	9.646	130	32.8	2.565

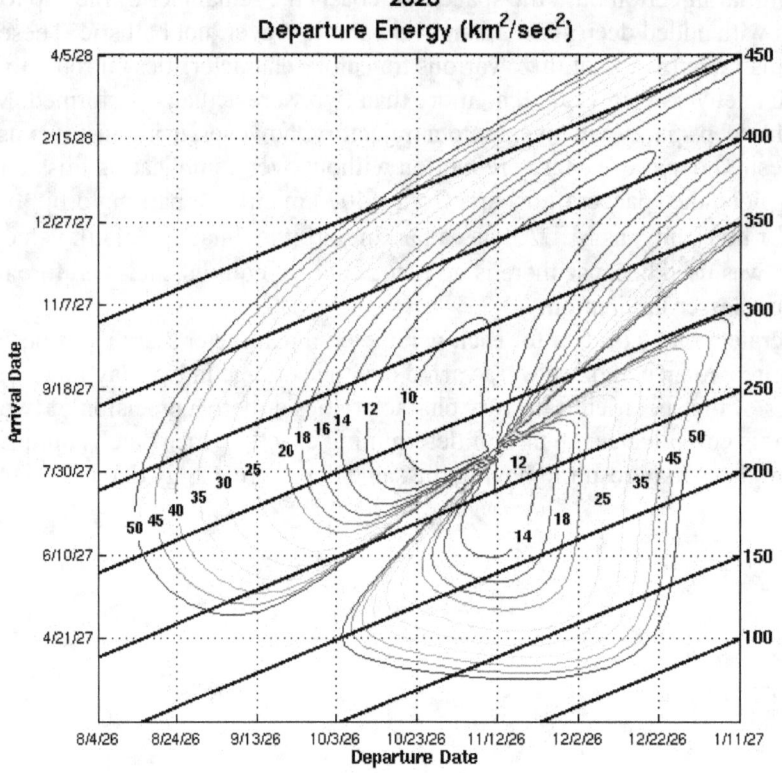

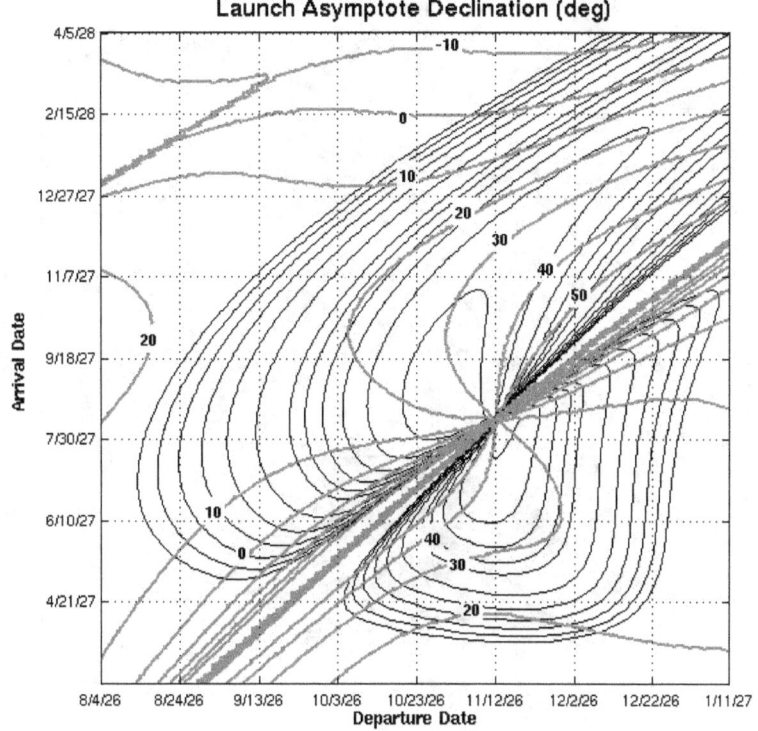

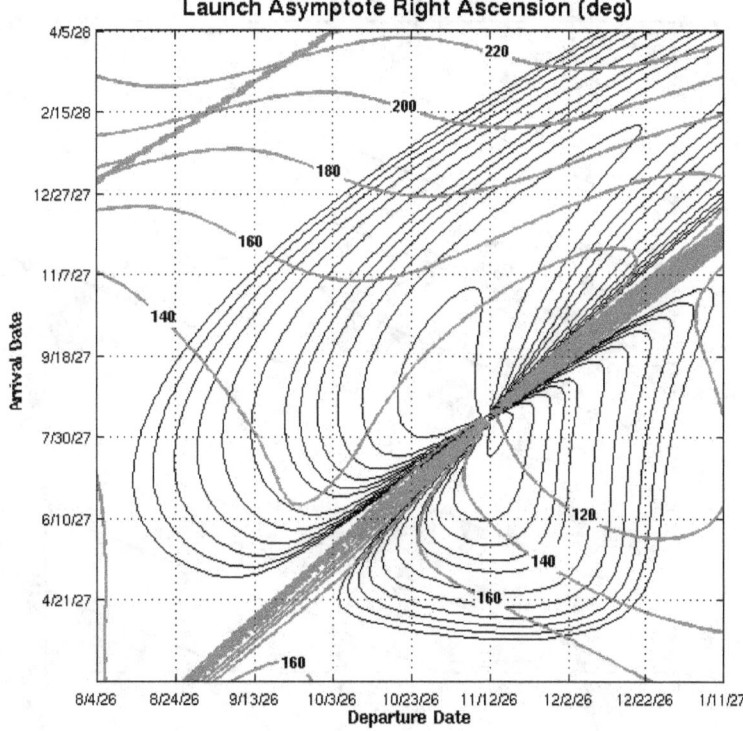

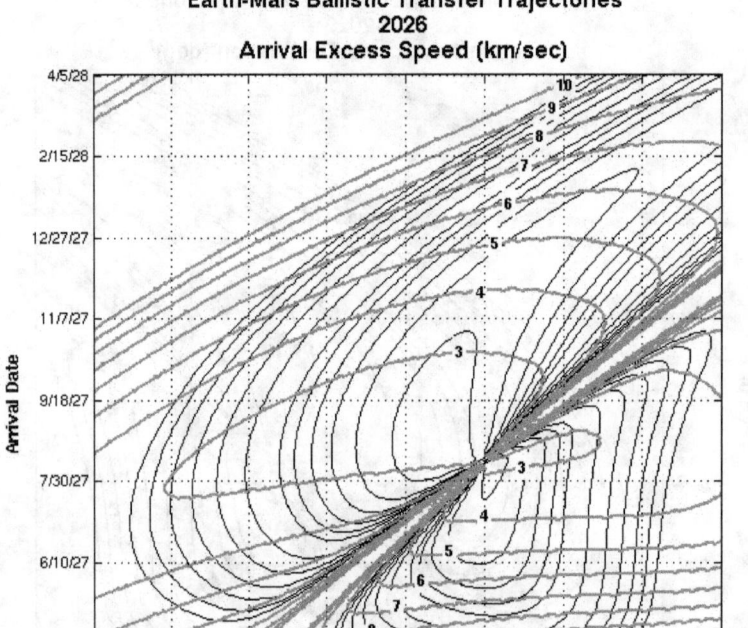

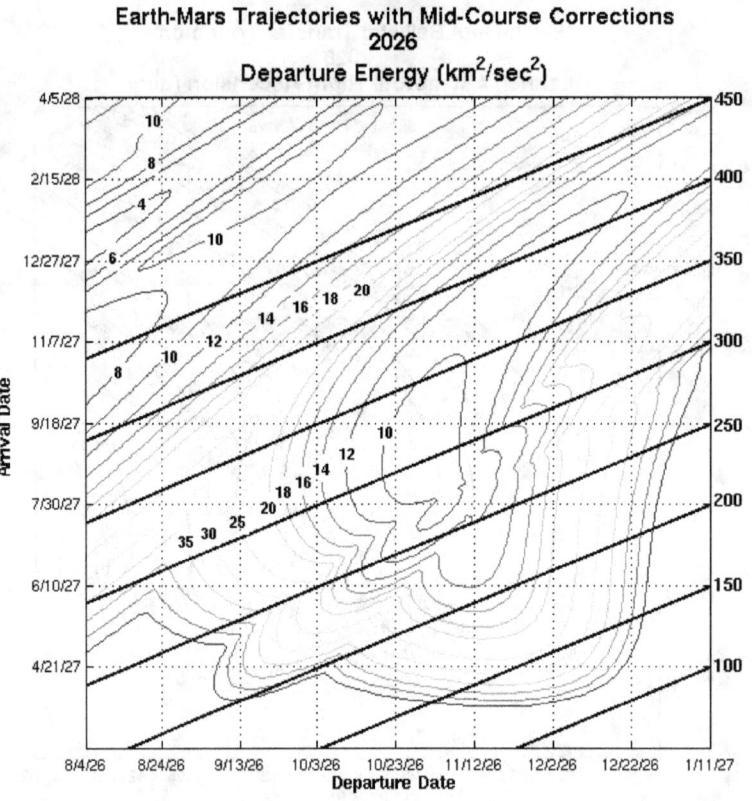

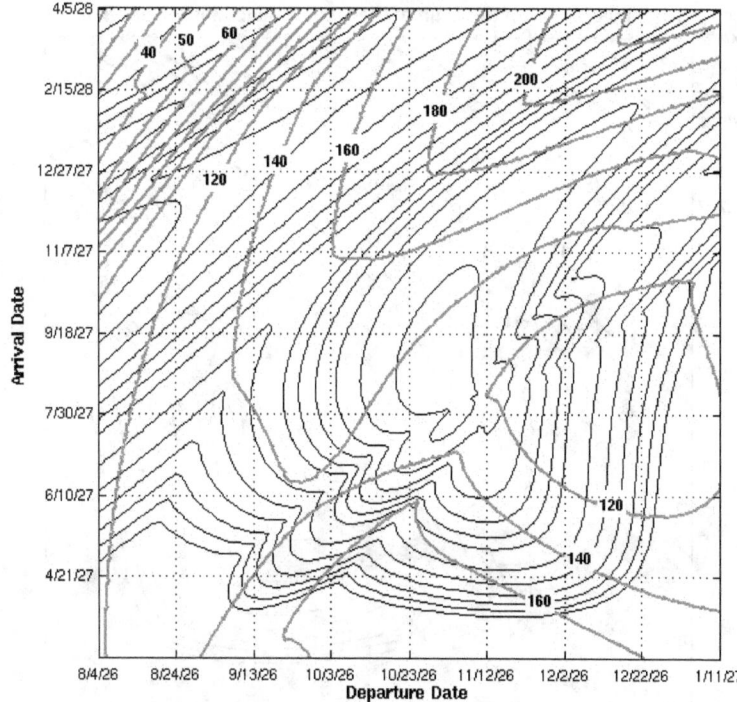

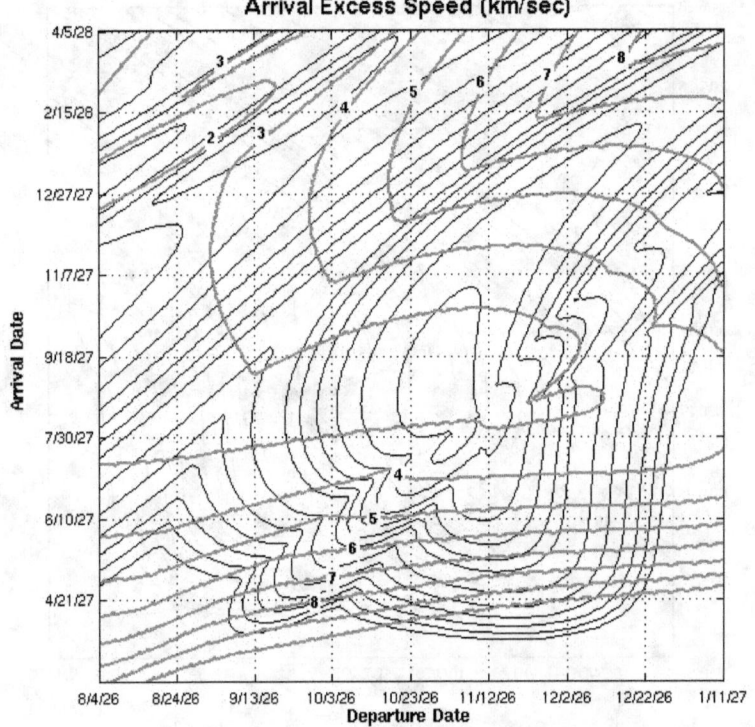
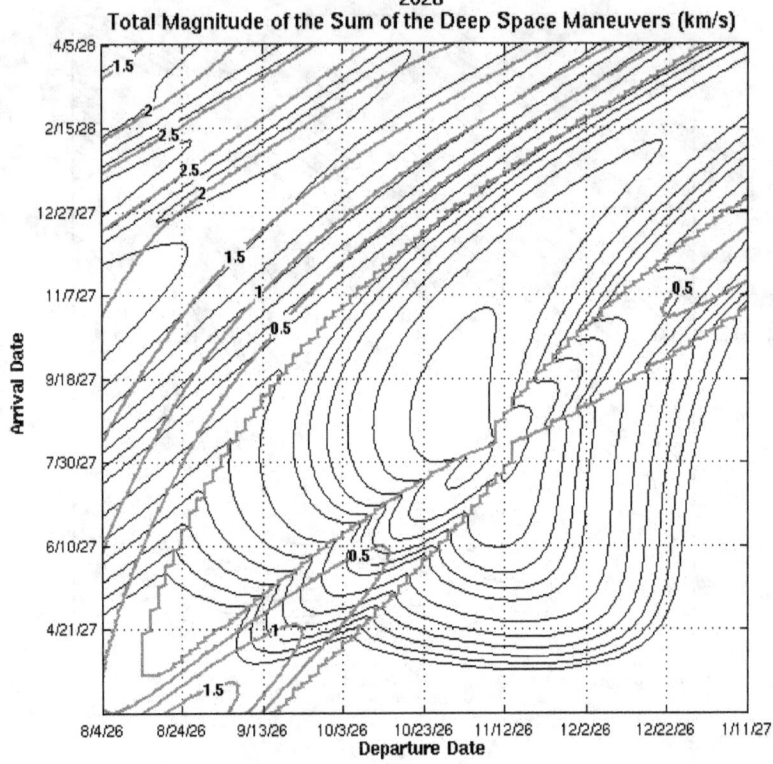

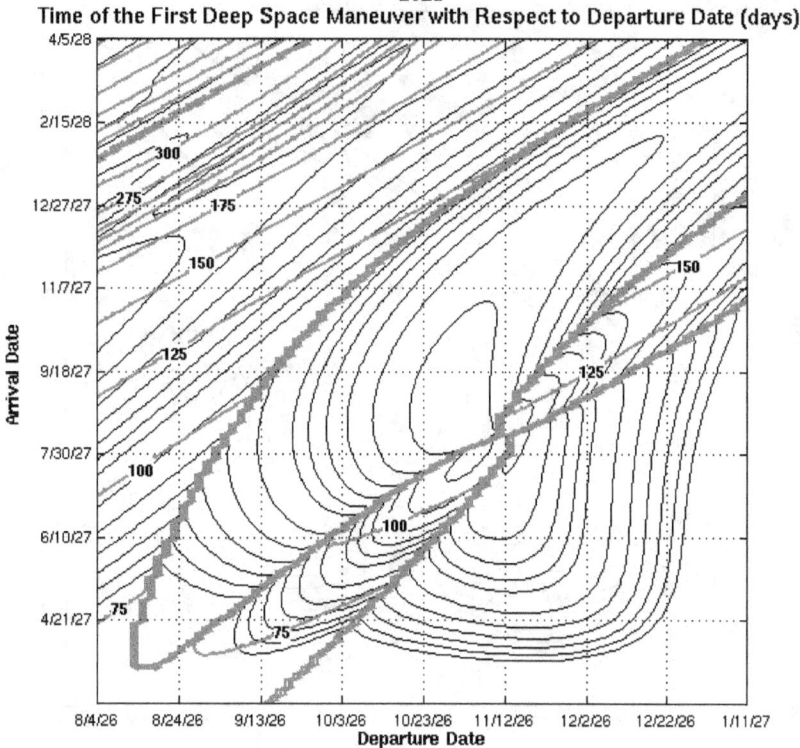

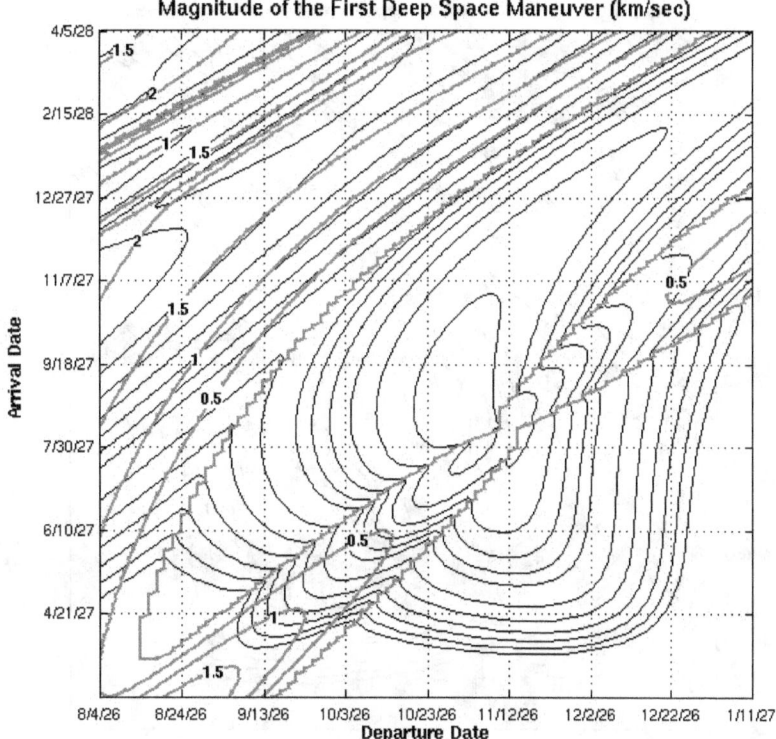

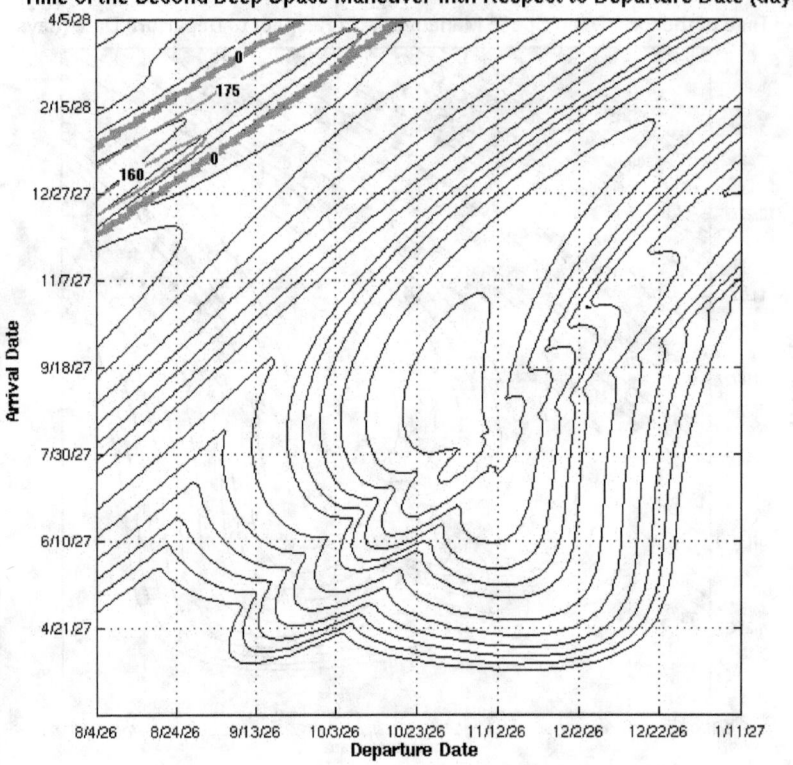

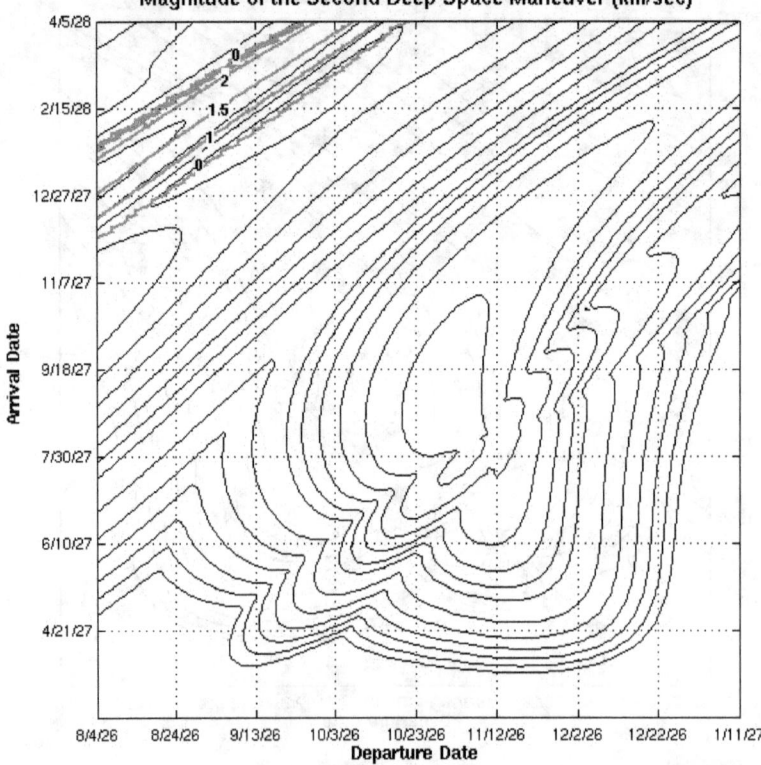

NASA/TM—2010-216764 18

Earth to Mars—2028 Opportunity

TABLE 3.—EARTH TO MARS—2028 OPPORTUNITY—ENERGY MINIMA

Mission type	Earth departure date (m/d/yr)	Mars arrival date (m/d/yr)	C_3 (km^2/sec^2)	Right ascension (deg)	Declination (deg)	Mars arrival excess speed (km/s)
Type 1	12/10/28	7/20/29	**9.048**	158.9	1.581	4.892
Type 2	12/2/28	10/16/29	**8.928**	185.1	29.34	3.261
Type 1	1/17/29	9/2/29	24.12	140.9	1.581	**3.593**
Type 2	11/20/28	9/18/29	9.315	182.8	25.51	**2.966**

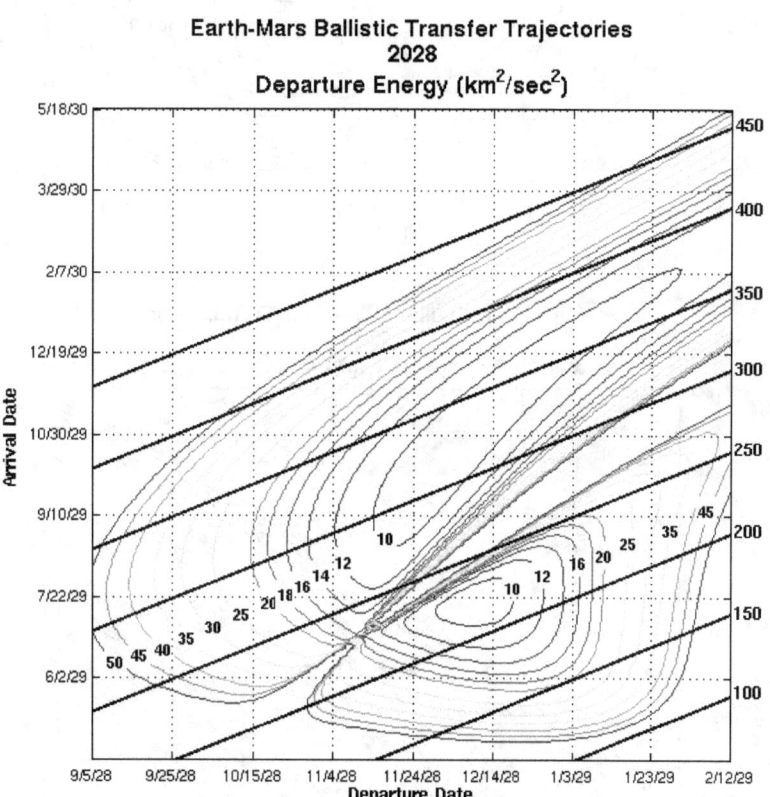

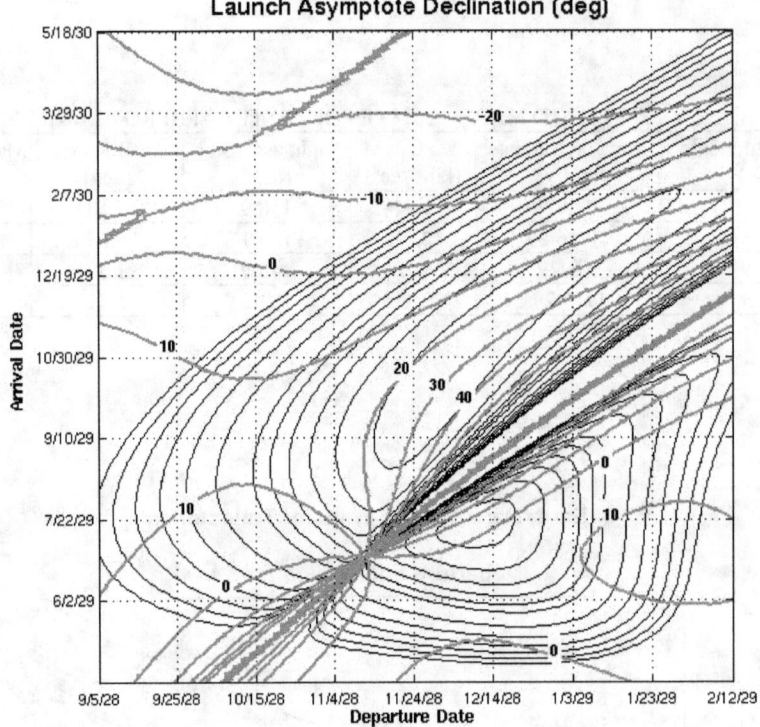

Earth-Mars Ballistic Transfer Trajectories
2028
Launch Asymptote Declination (deg)

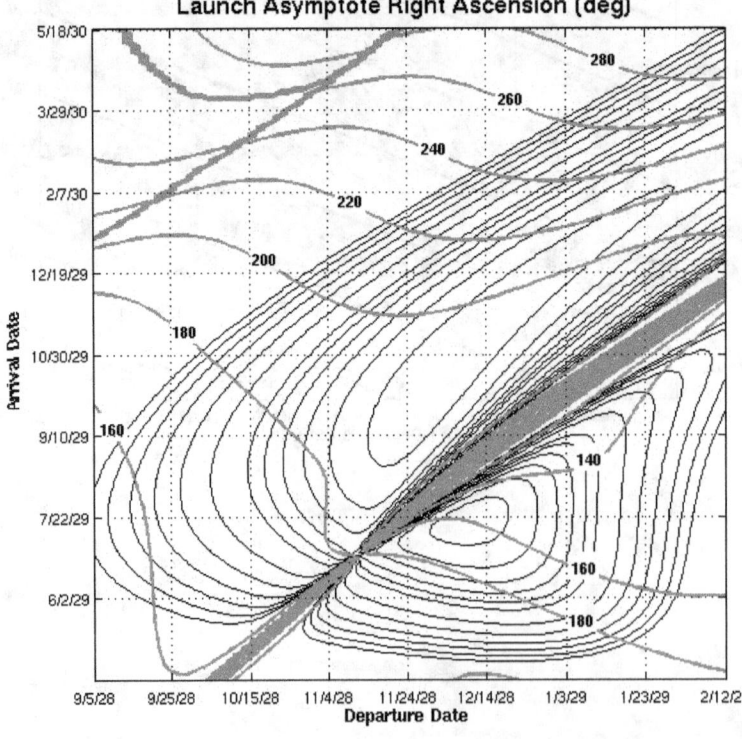

Earth-Mars Ballistic Transfer Trajectories
2028
Launch Asymptote Right Ascension (deg)

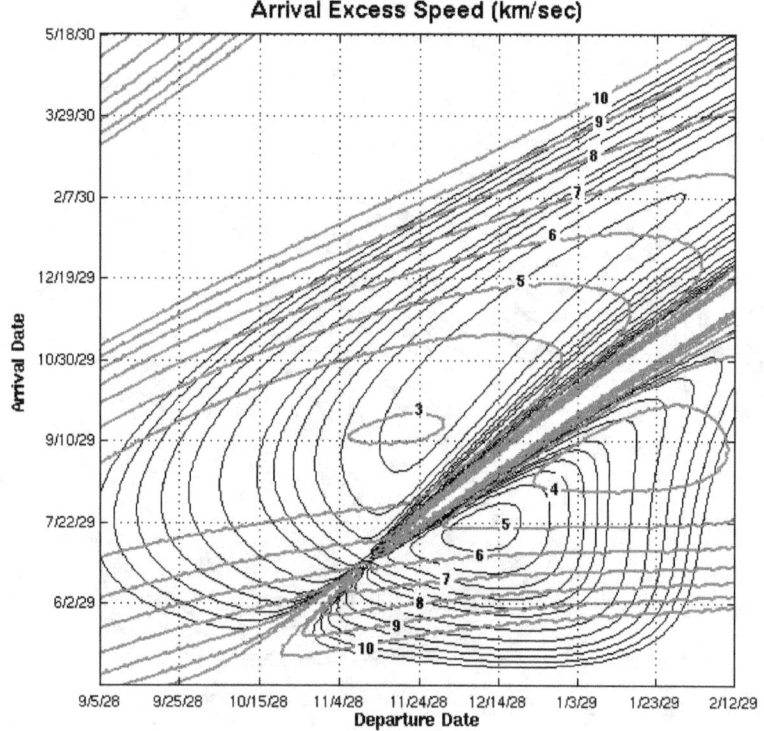

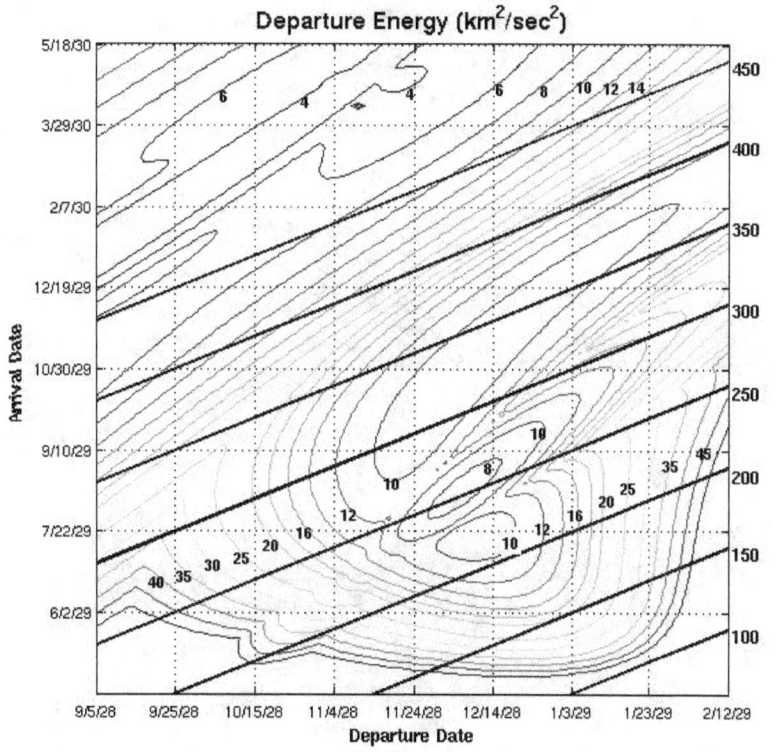

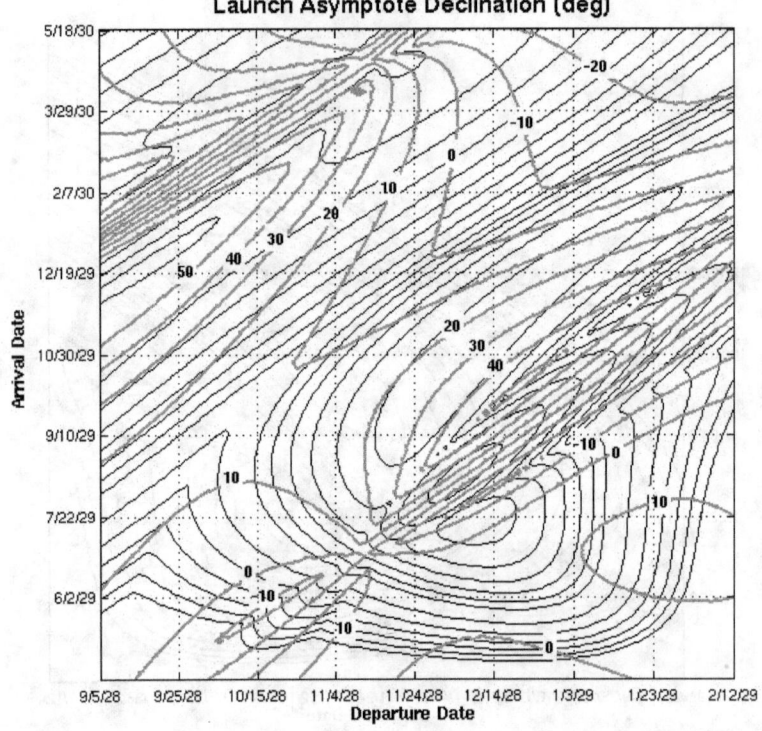

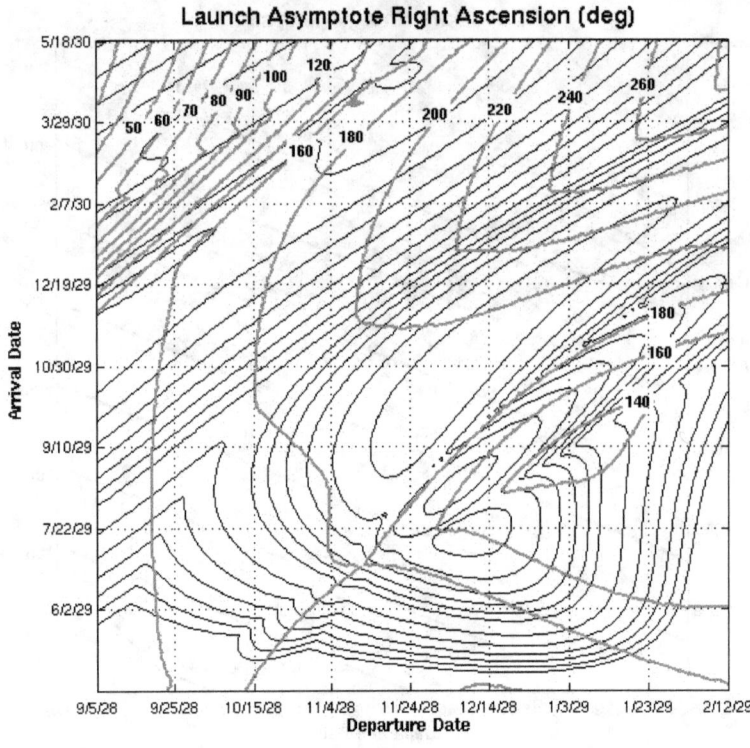

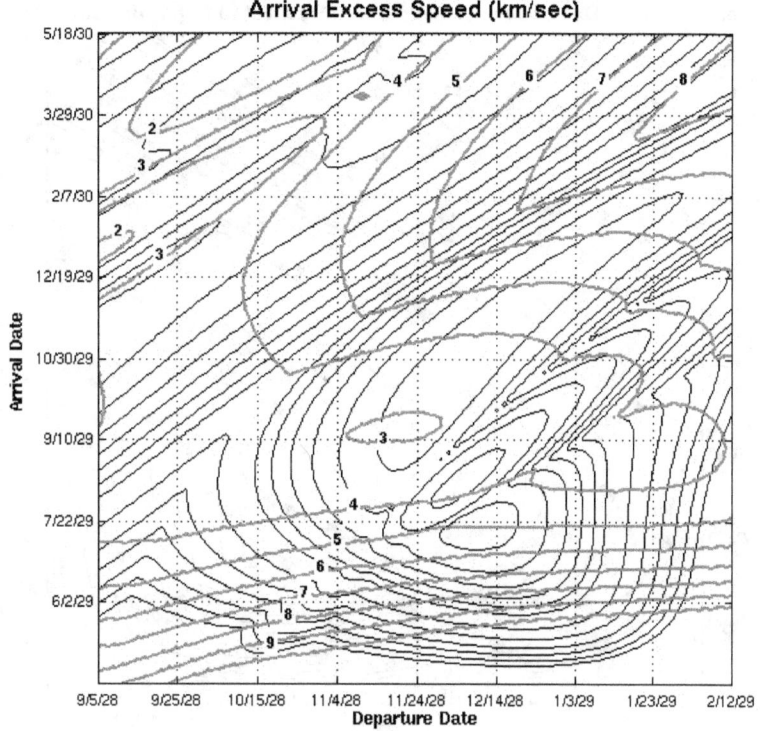
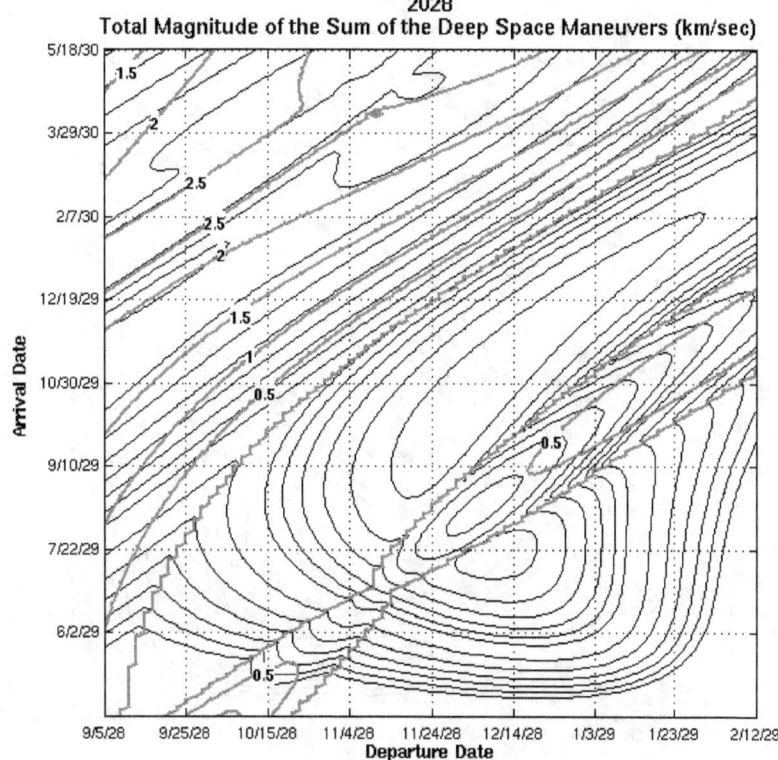

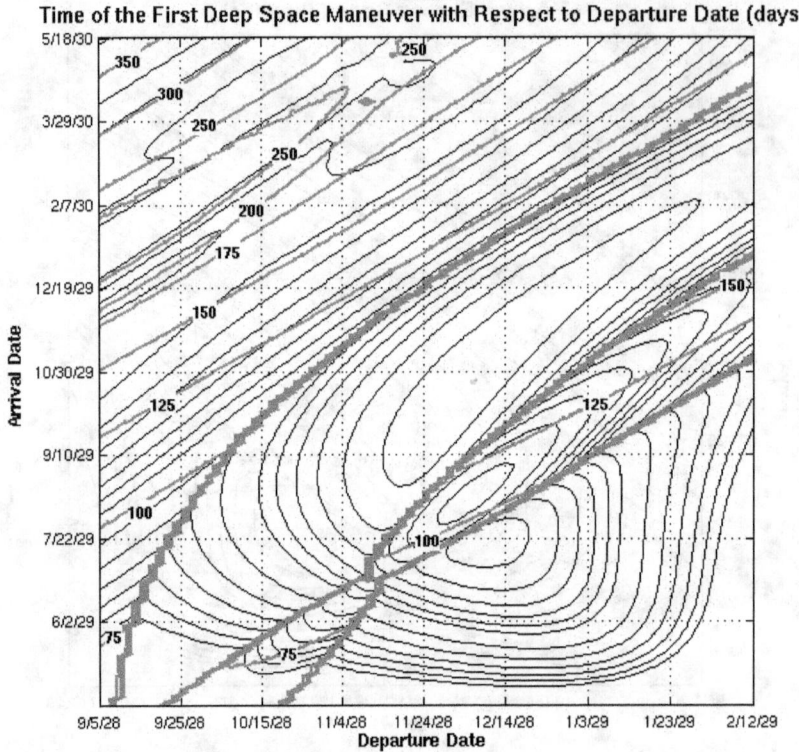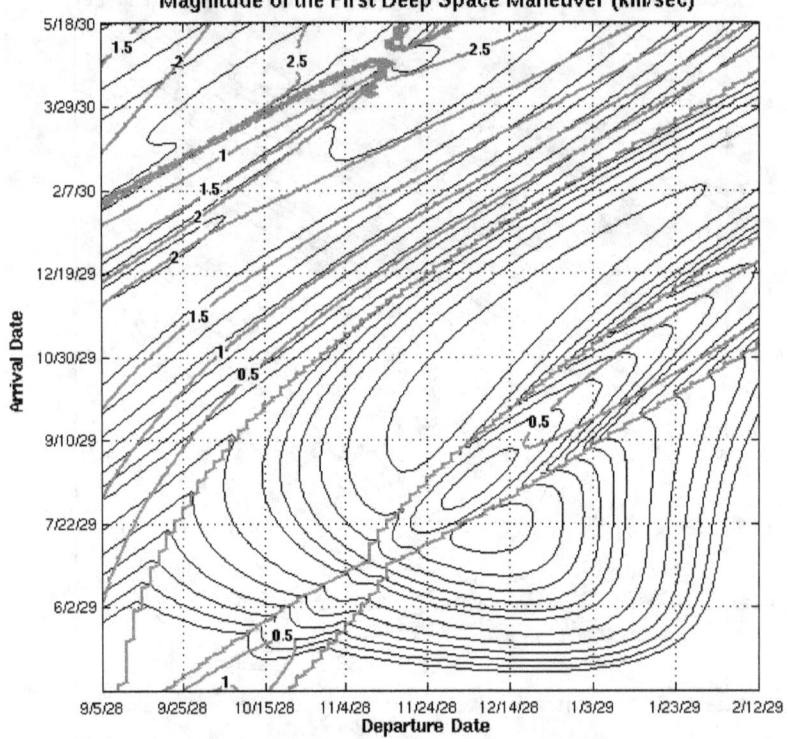

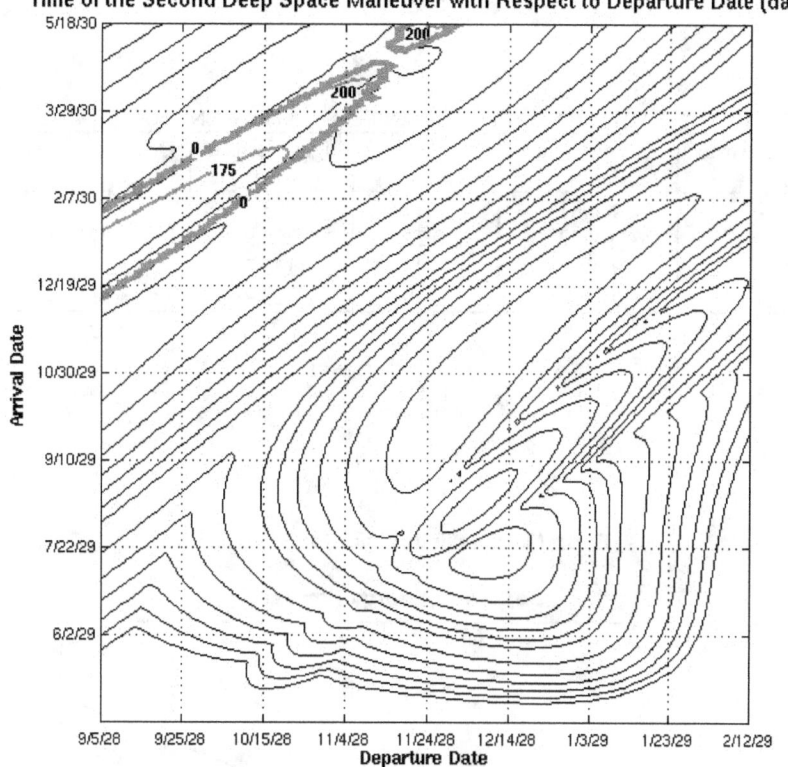
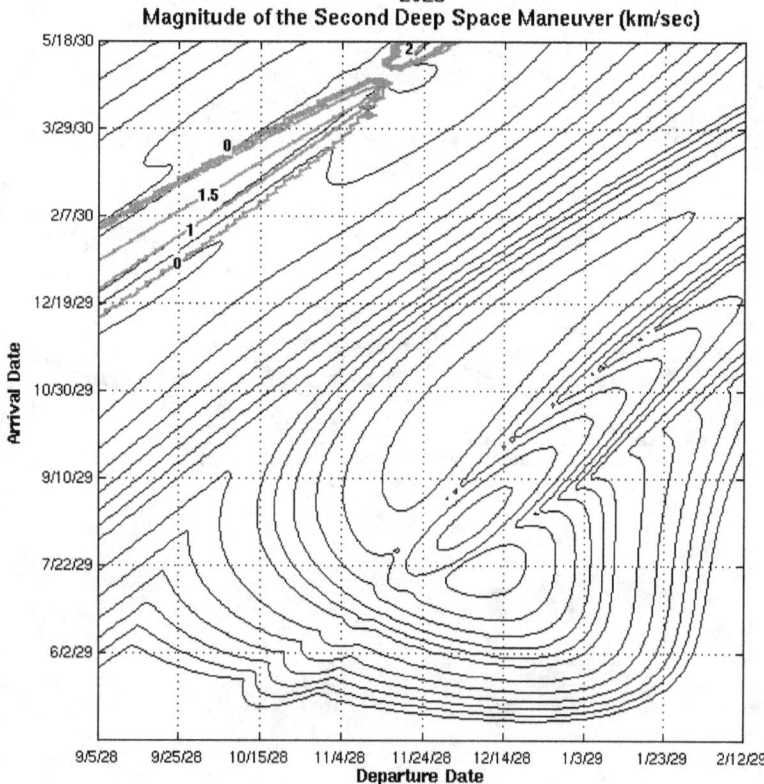

Earth to Mars—2031 Opportunity

TABLE 4.—EARTH TO MARS—2031 OPPORTUNITY—ENERGY MINIMA

Mission type	Earth departure date (m/d/yr)	Mars arrival date (m/d/yr)	C_3 (km^2/sec^2)	Right ascension (deg)	Declination (deg)	Mars arrival excess speed (km/s)
Type 1	1/28/31	8/6/31	**9.00**	193.8	−34.2	5.541
Type 2	2/23/31	1/9/32	**8.237**	252.9	1.015	5.53
Type 1	3/1/31	9/27/31	17.89	177.7	−25.3	**3.777**
Type 2	12/13/30	9/25/31	12.48	225.9	8.543	**3.445**

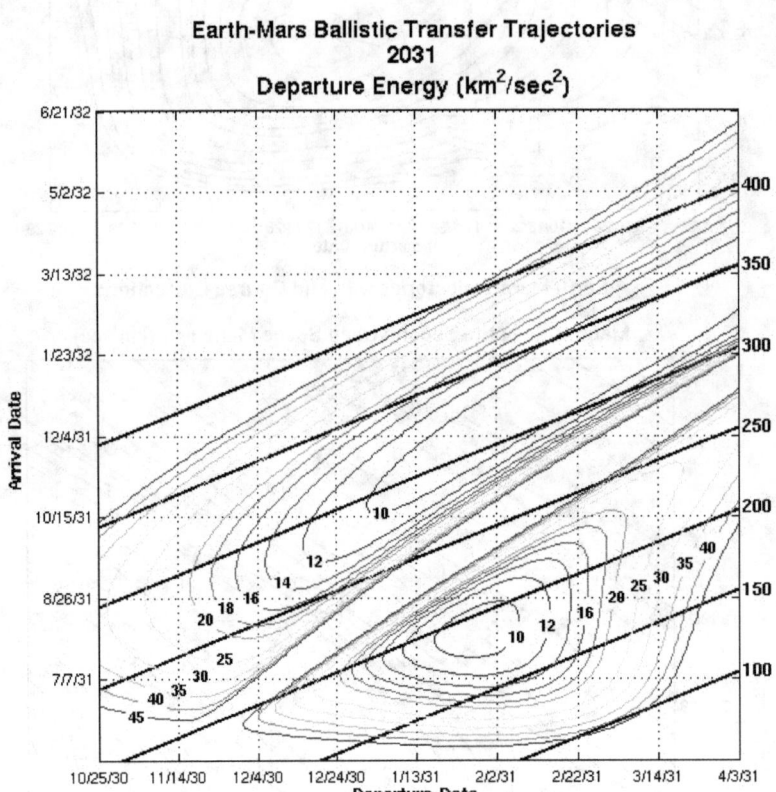

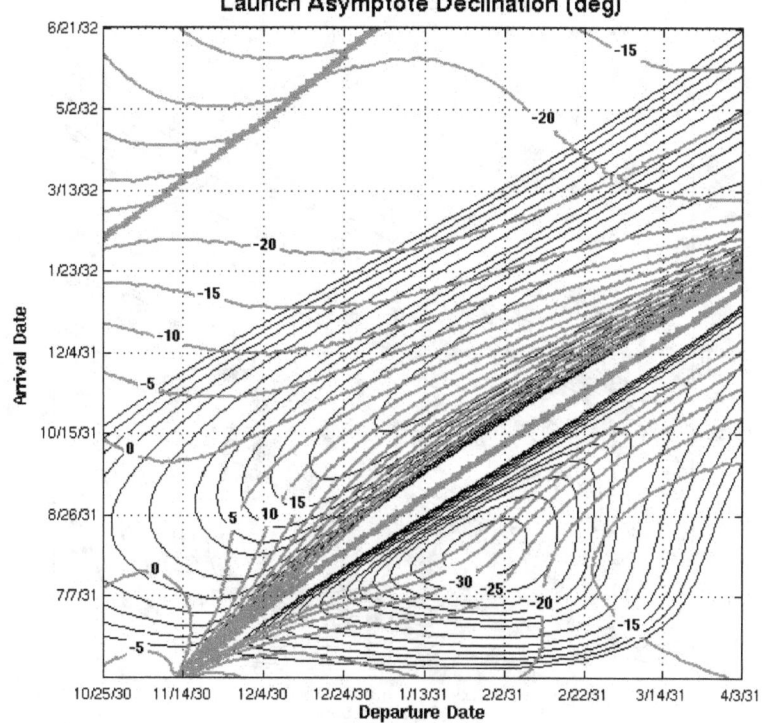

Earth-Mars Ballistic Transfer Trajectories
2031
Launch Asymptote Declination (deg)

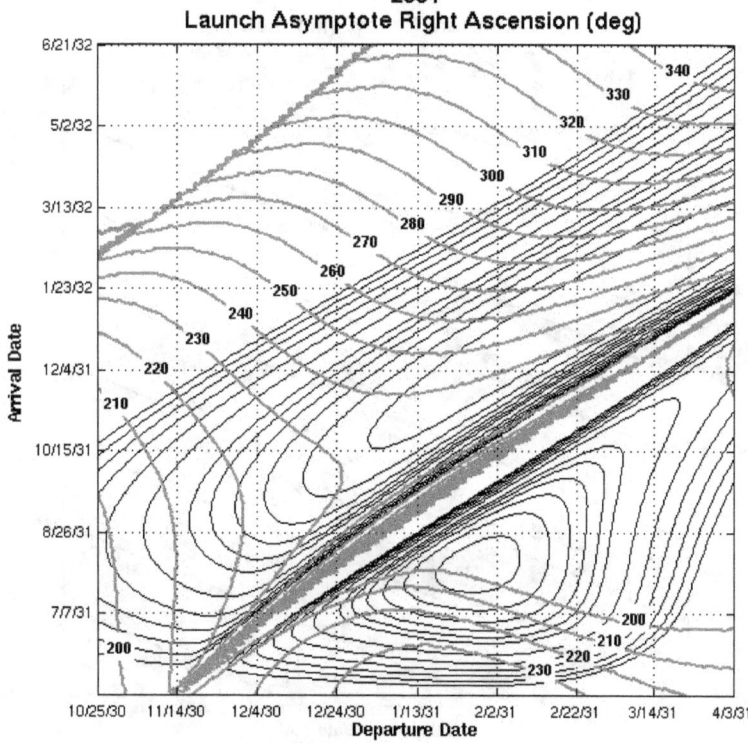

Earth-Mars Ballistic Transfer Trajectories
2031
Launch Asymptote Right Ascension (deg)

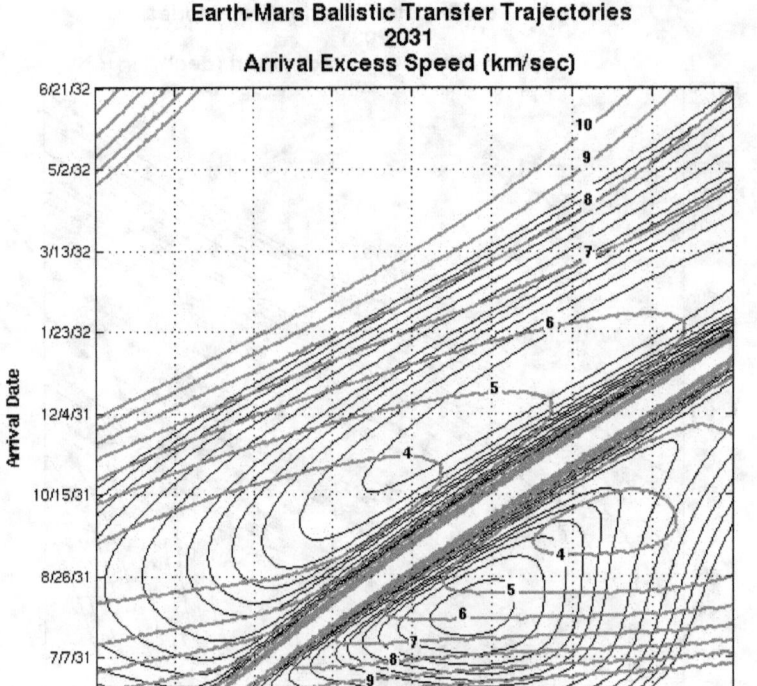

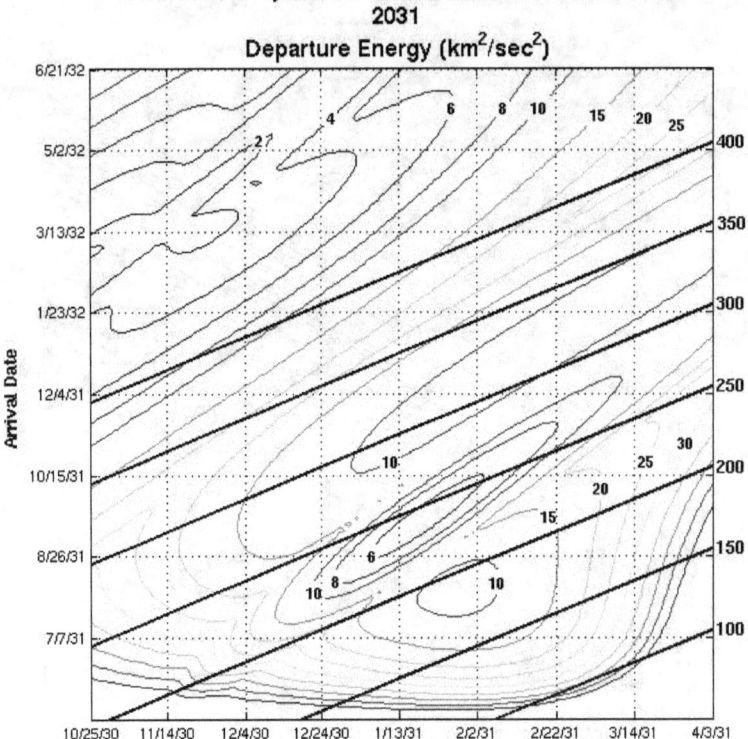

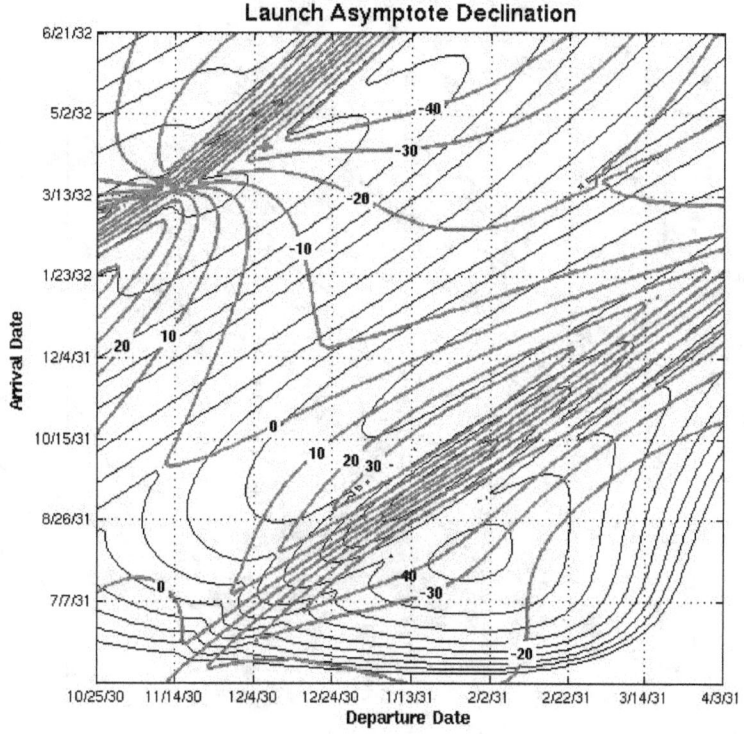

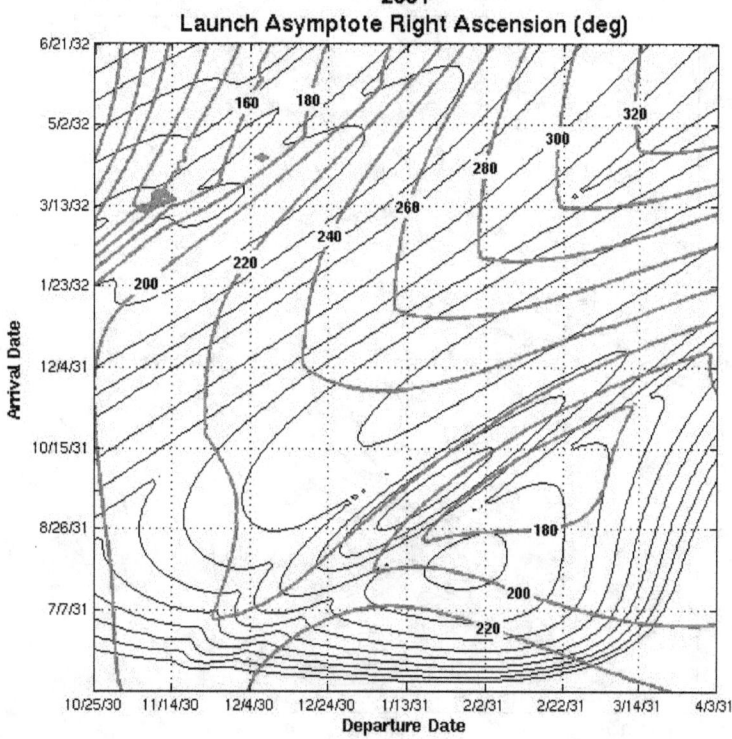

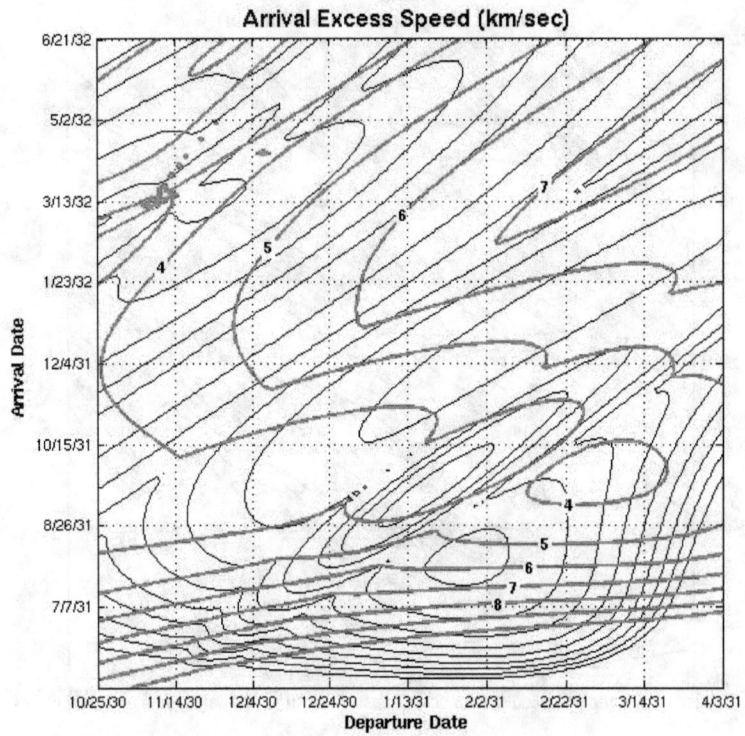

Earth-Mars Trajectories with Mid-Course Corrections
2031
Arrival Excess Speed (km/sec)

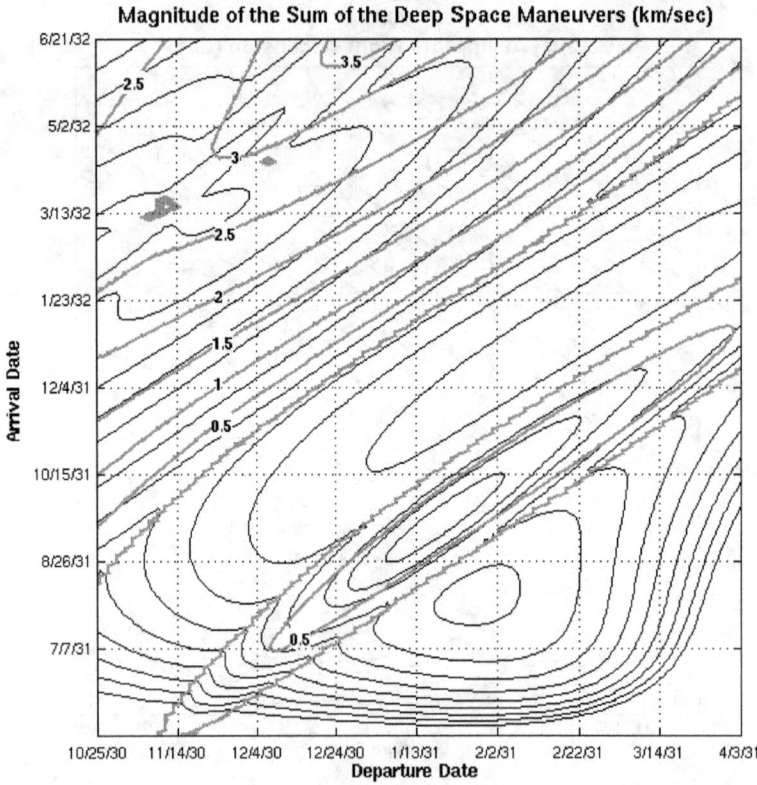

Earth-Mars Trajectories with Mid-Course Corrections
2031
Magnitude of the Sum of the Deep Space Maneuvers (km/sec)

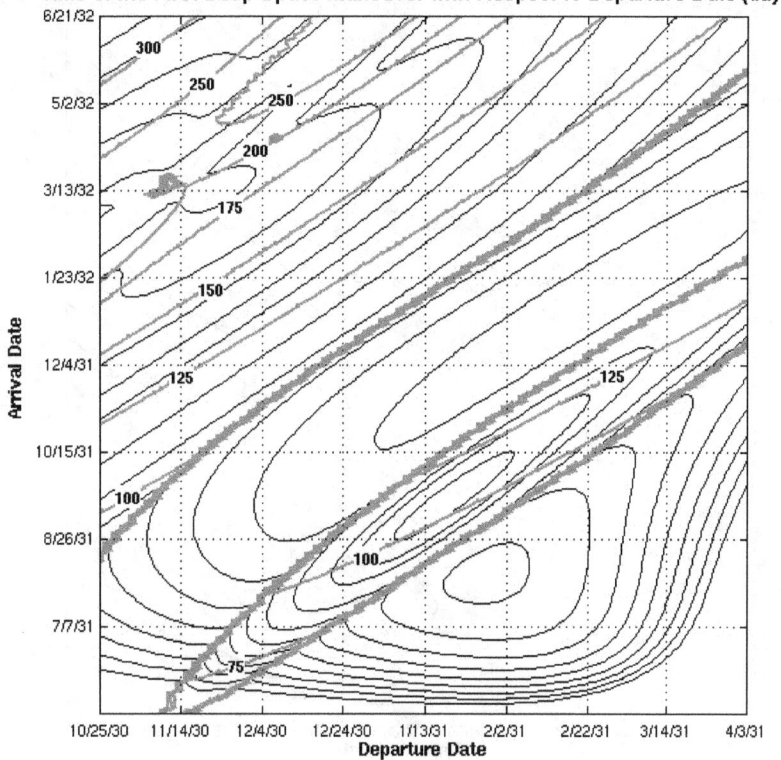

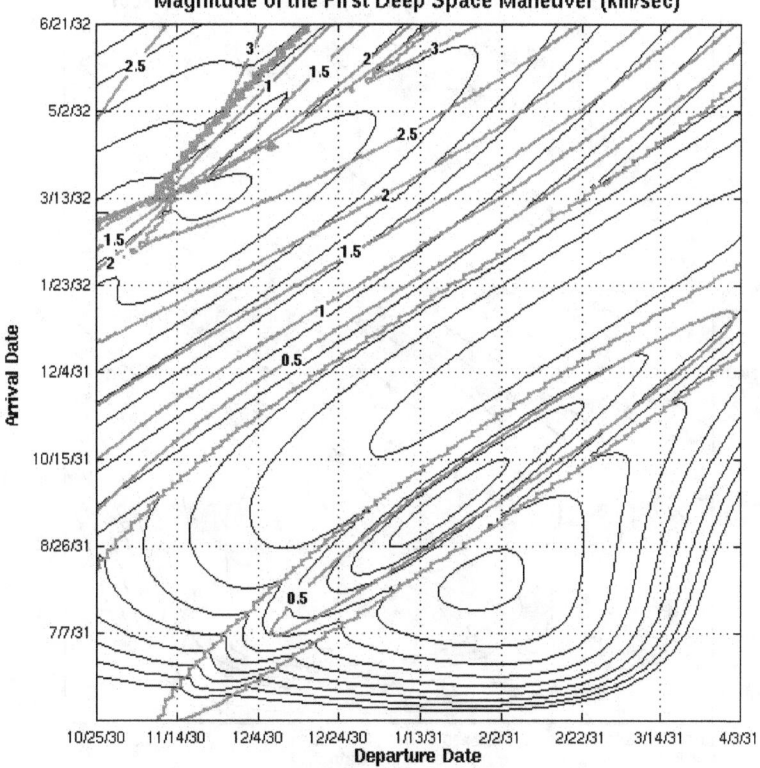

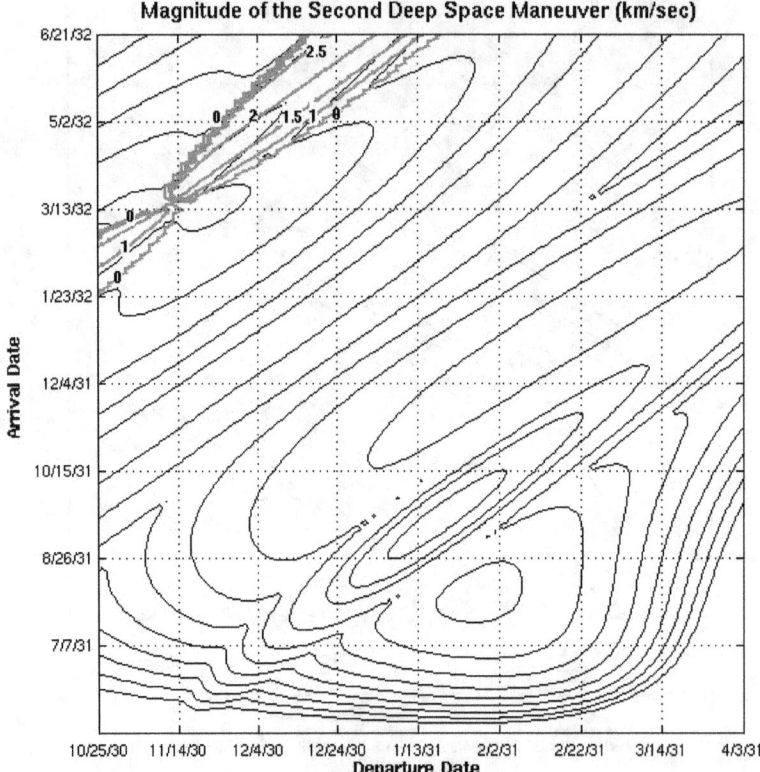

Earth to Mars—2033 Opportunity

TABLE 5.—EARTH TO MARS—2033 OPPORTUNITY—ENERGY MINIMA

Mission type	Earth departure date (m/d/yr)	Mars arrival date (m/d/yr)	C_3 (km^2/sec^2)	Right ascension (deg)	Declination (deg)	Mars arrival excess speed (km/s)
Type 1	4/6/33	10/1/33	8.412	271	−54.9	3.956
Type 2	4/28/33	1/27/34	7.781	311.4	−11.2	4.377
Type 1	4/20/33	11/6/33	9.266	267.1	−53.2	3.311
Type 2	1/26/33	10/17/33	17.78	278.3	−2.53	3.831

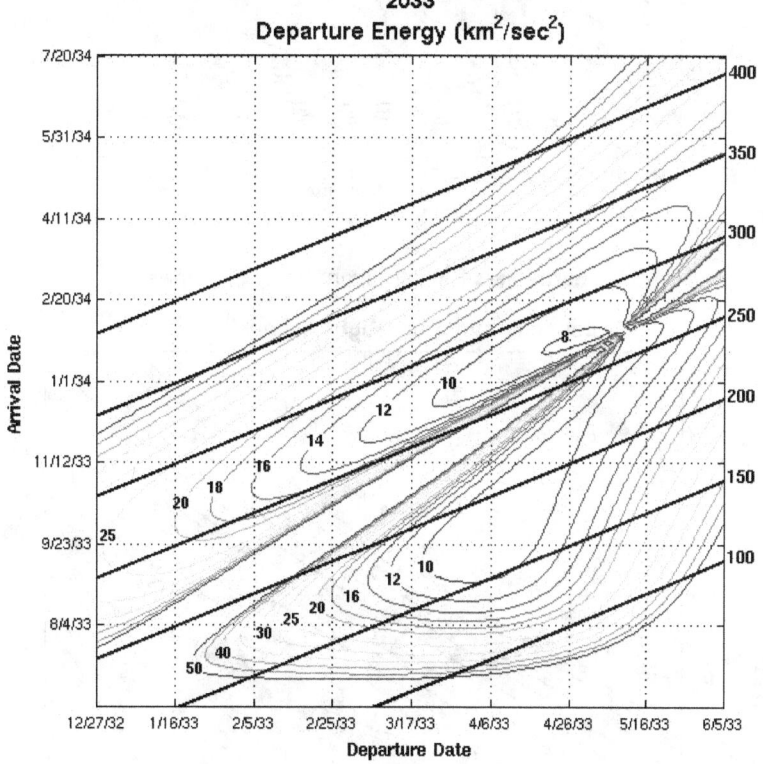

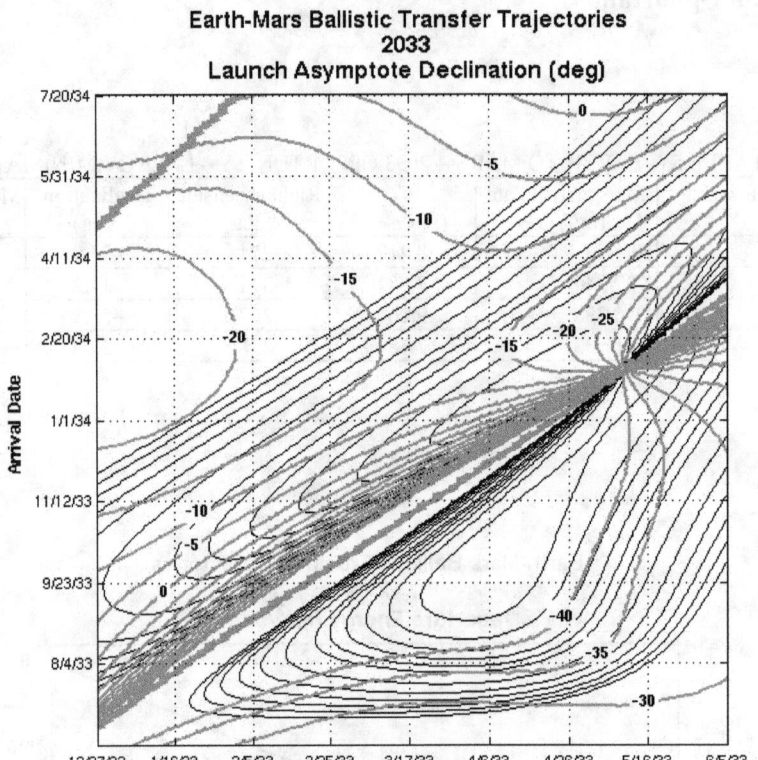

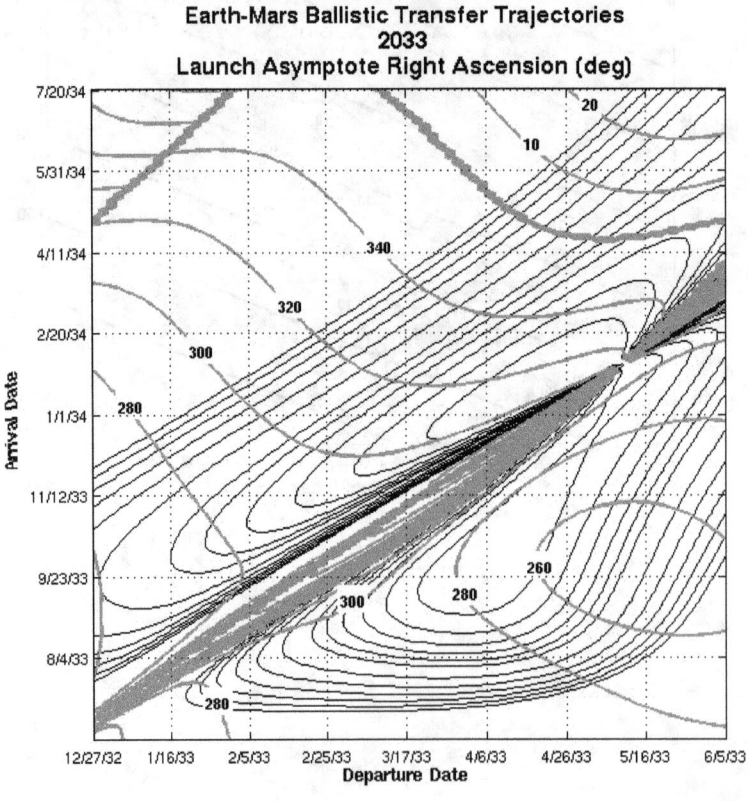

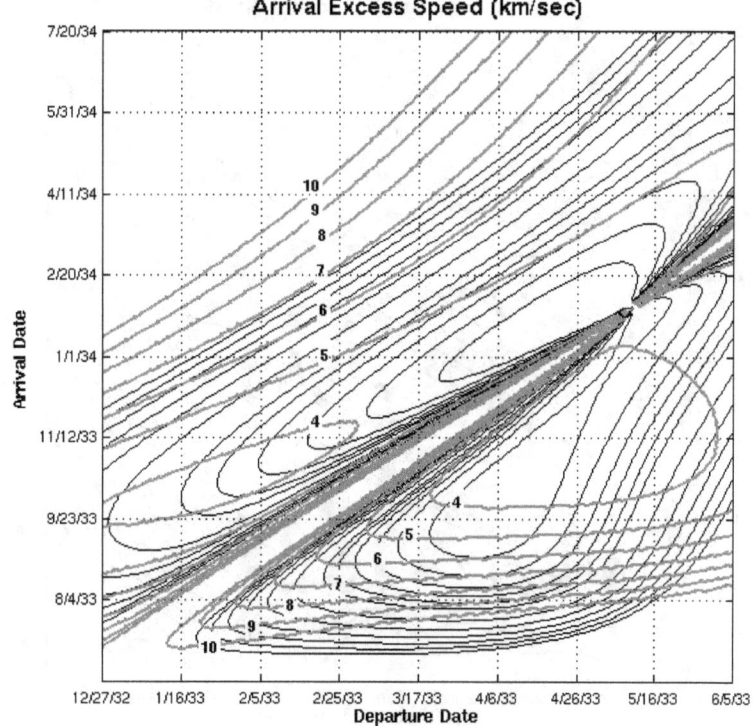

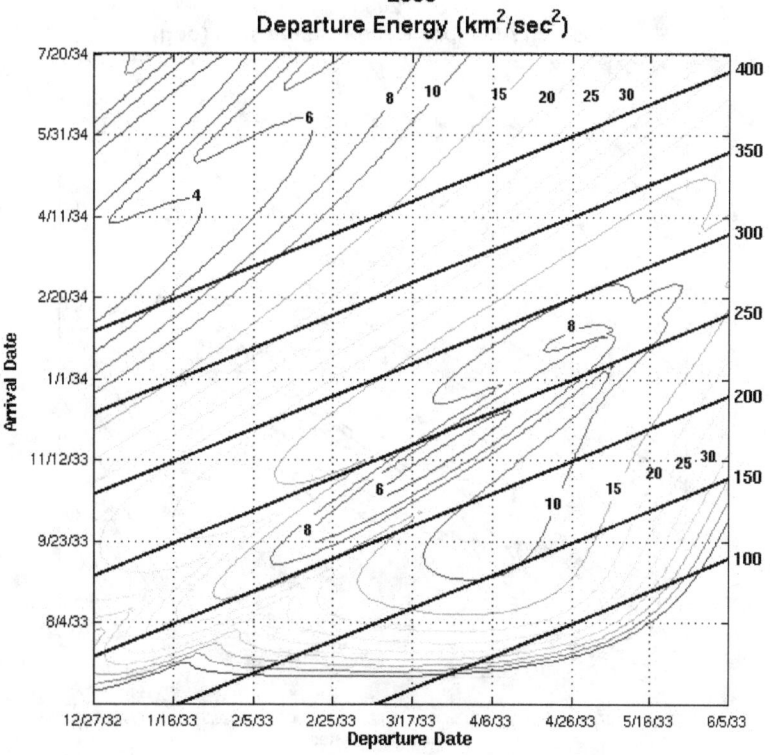

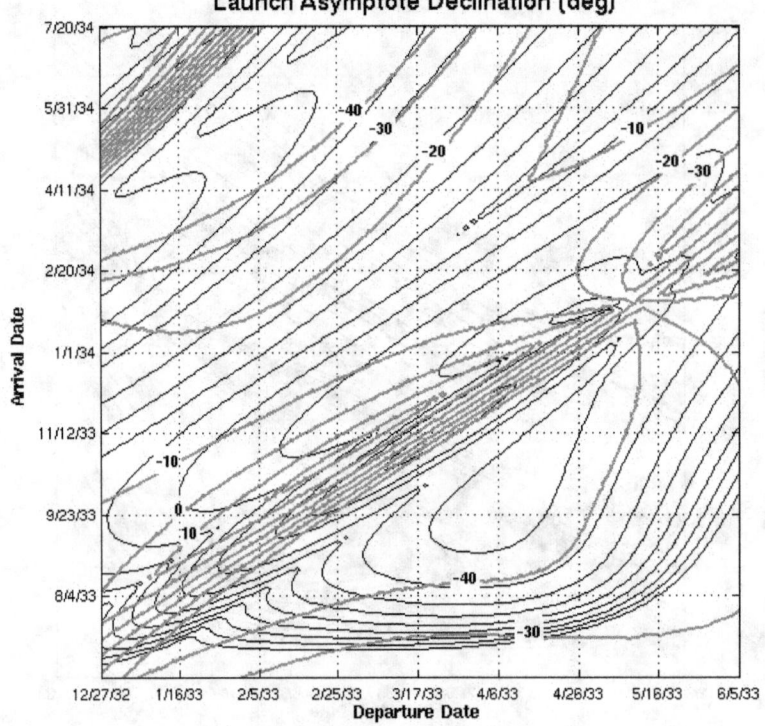

Earth-Mars Trajectories with Mid-Course Corrections
2033
Launch Asymptote Declination (deg)

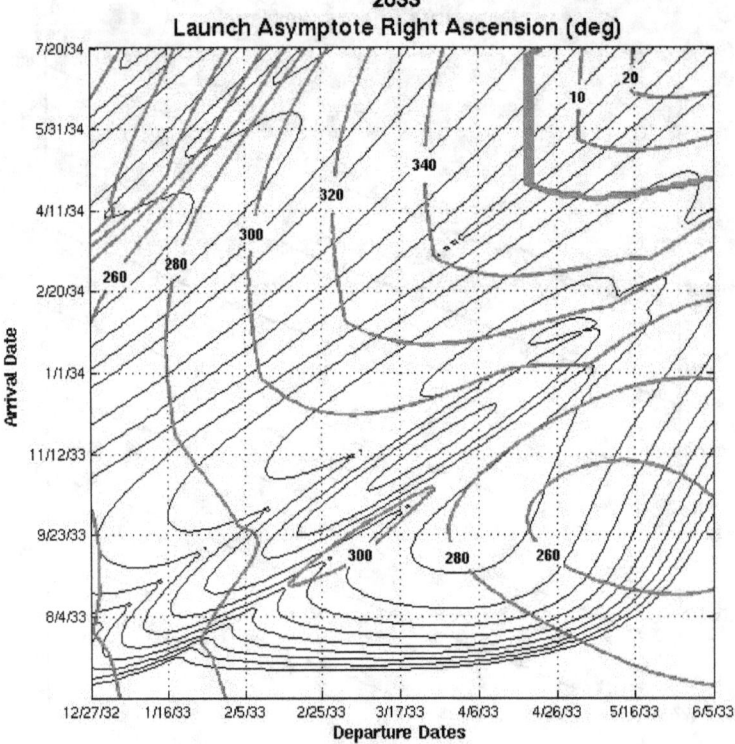

Earth-Mars Trajectories with Mid-Course Corrections
2033
Launch Asymptote Right Ascension (deg)

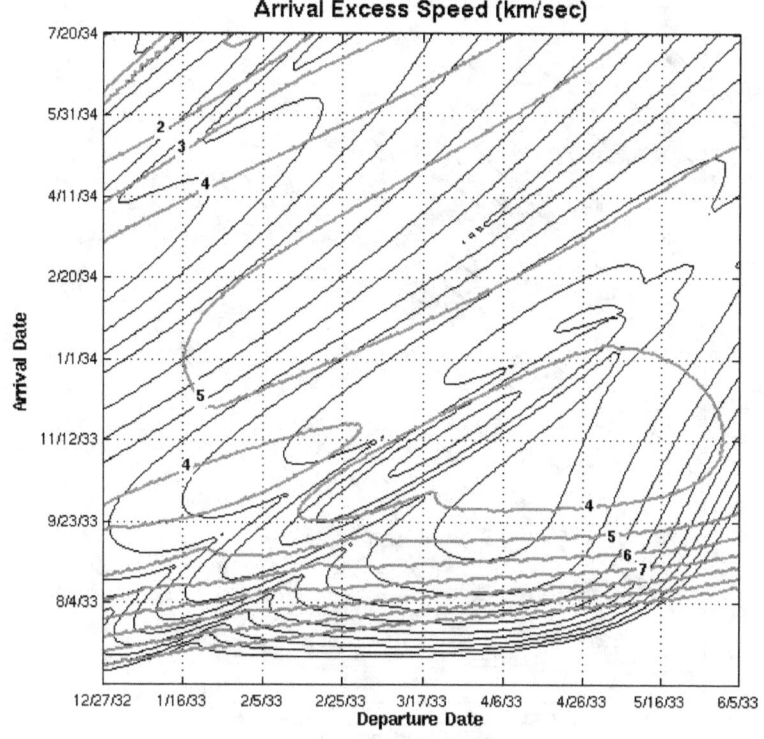
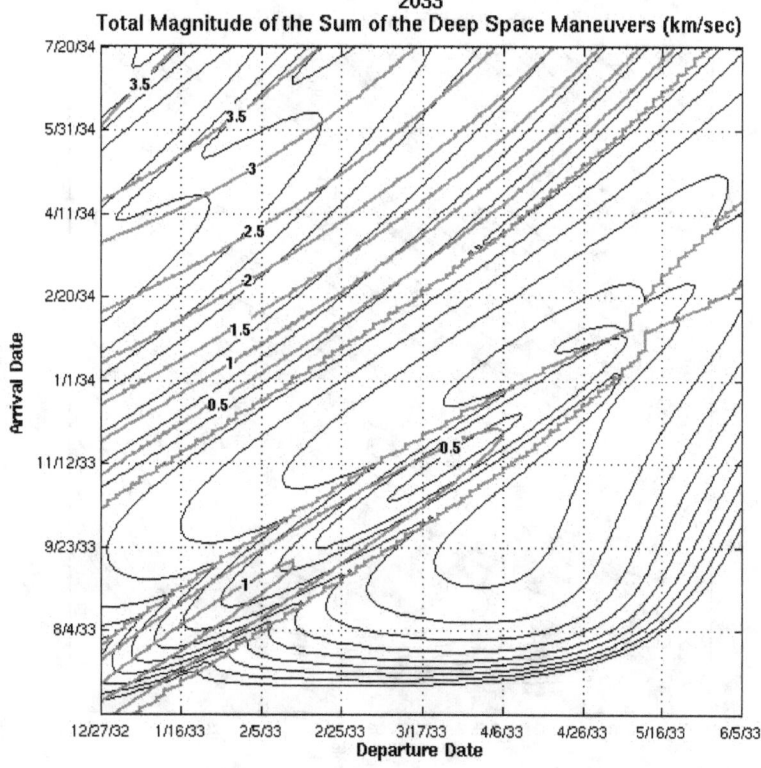

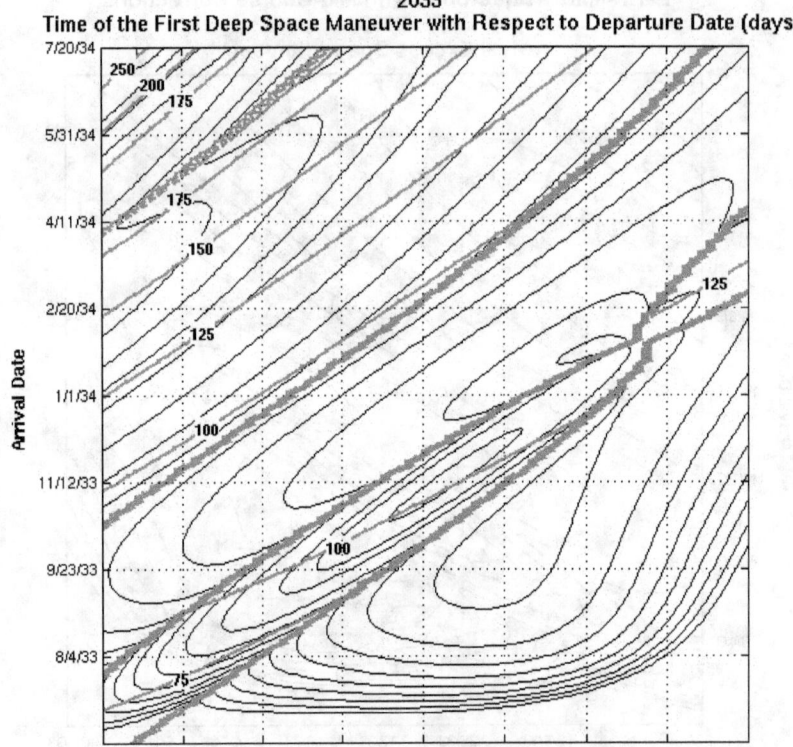

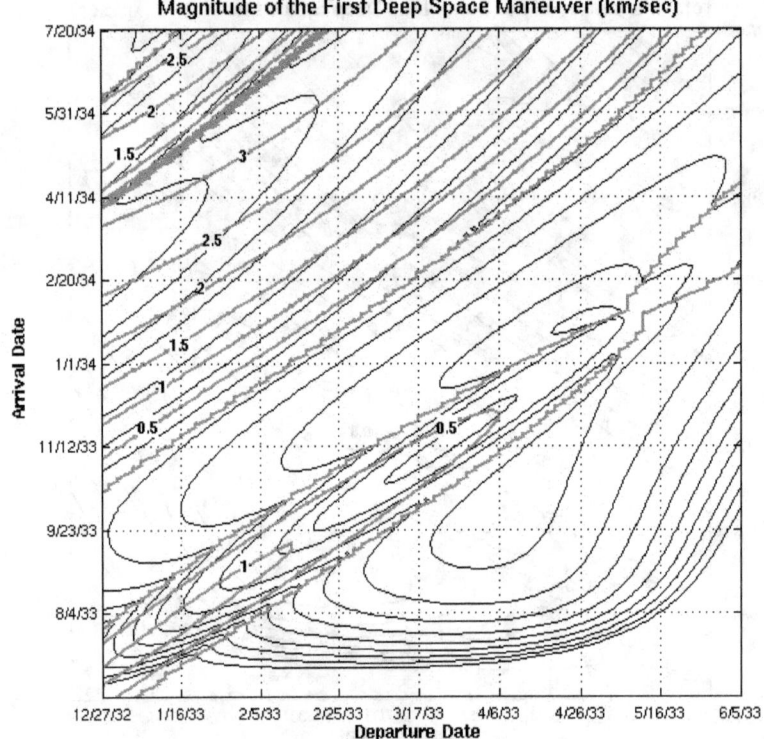

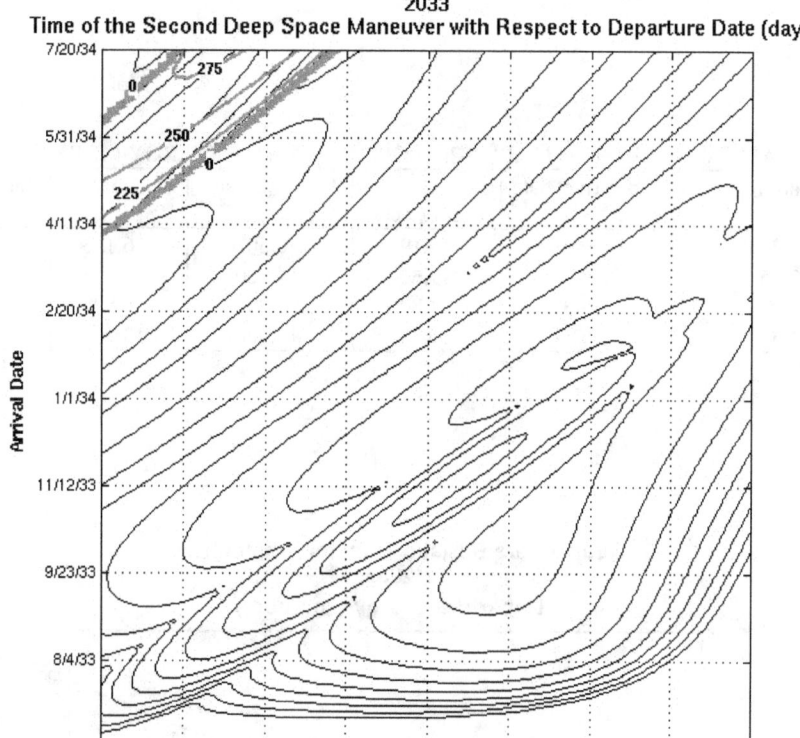

Earth-Mars Trajectories with Mid-Course Corrections
2033
Time of the Second Deep Space Maneuver with Respect to Departure Date (days)

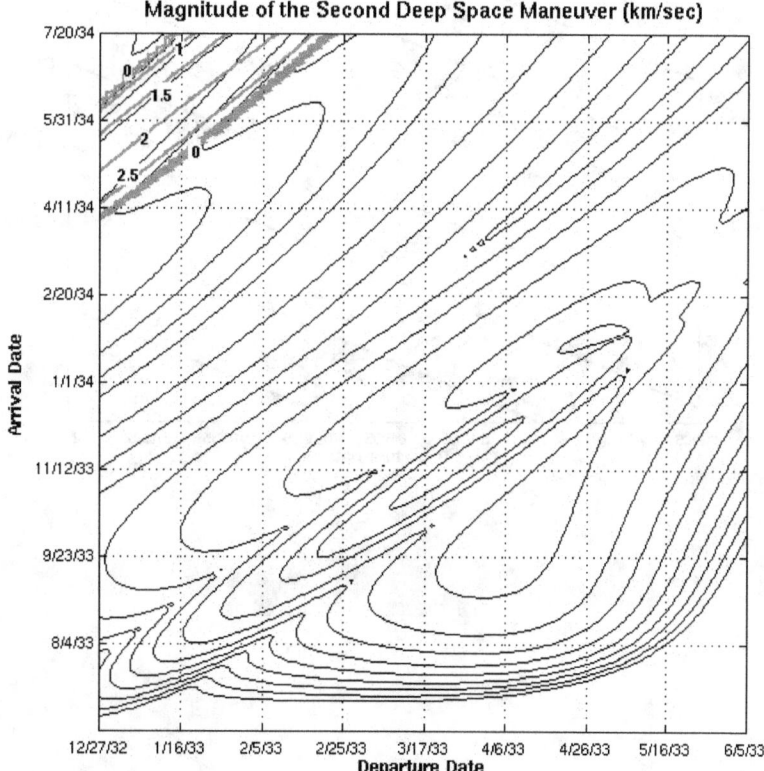

Earth-Mars Trajectories with Mid-Course Corrections
2033
Magnitude of the Second Deep Space Maneuver (km/sec)

Earth to Mars—2035 Opportunity

TABLE 6.—EARTH TO MARS—2035 OPPORTUNITY—ENERGY MINIMA

Mission type	Earth departure date (m/d/yr)	Mars arrival date (m/d/yr)	C_3 (km²/sec²)	Right ascension (deg)	Declination (deg)	Mars arrival excess speed (km/s)
Type 1	4/21/35	11/3/35	**10.19**	0.5033	6.135	2.692
Type 2	6/12/35	7/28/36	**17.52**	54.360	2.953	4.063
Type 1	5/7/35	11/23/35	11.80	351.300	6.892	**2.600**
Type 2	3/12/35	11/7/35	19.33	17.900	−13.50	**2.697**

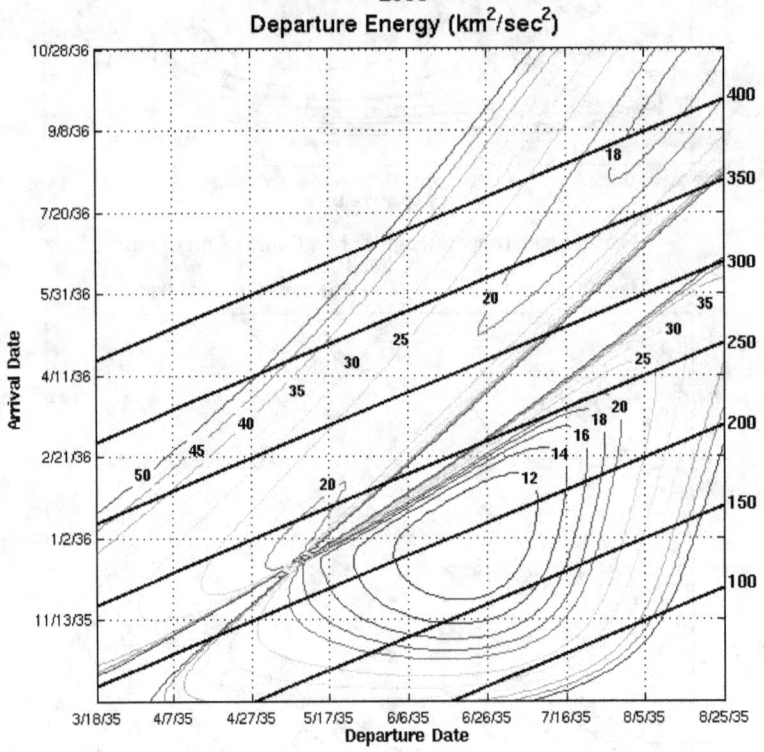

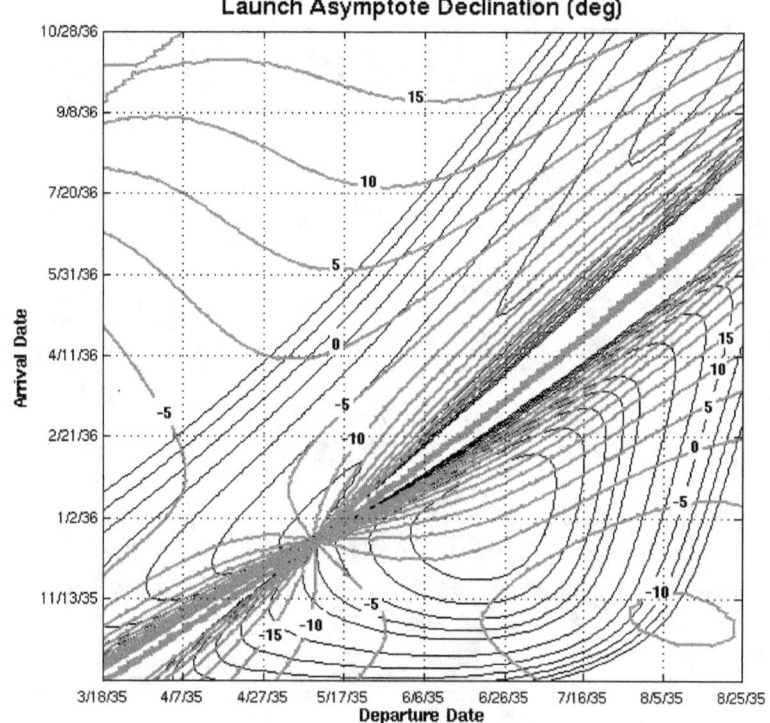

Earth-Mars Ballistic Transfer Trajectories
2035
Launch Asymptote Declination (deg)

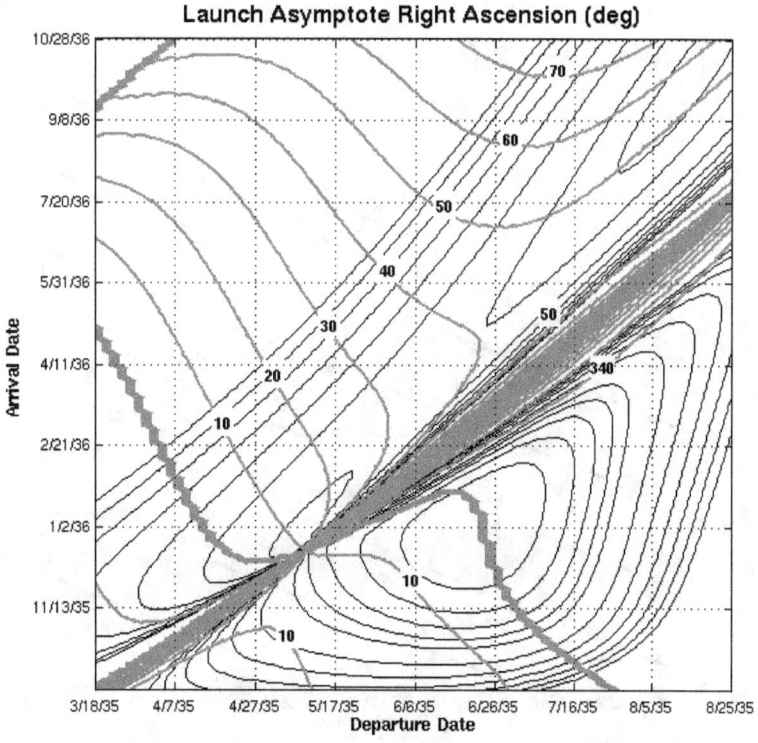

Earth-Mars Ballistic Transfer Trajectories
2035
Launch Asymptote Right Ascension (deg)

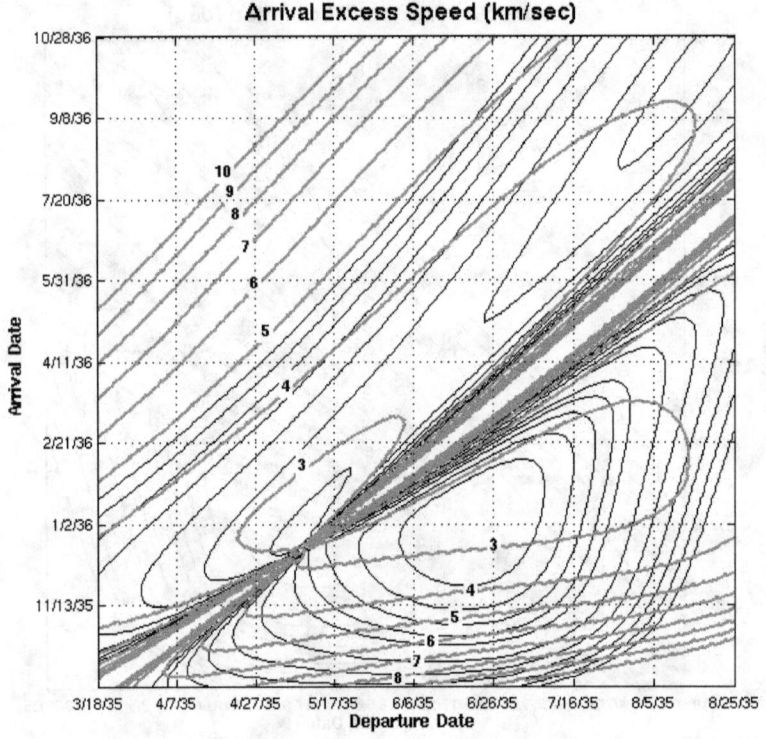

Earth-Mars Ballistic Transfer Trajectories
2035
Arrival Excess Speed (km/sec)

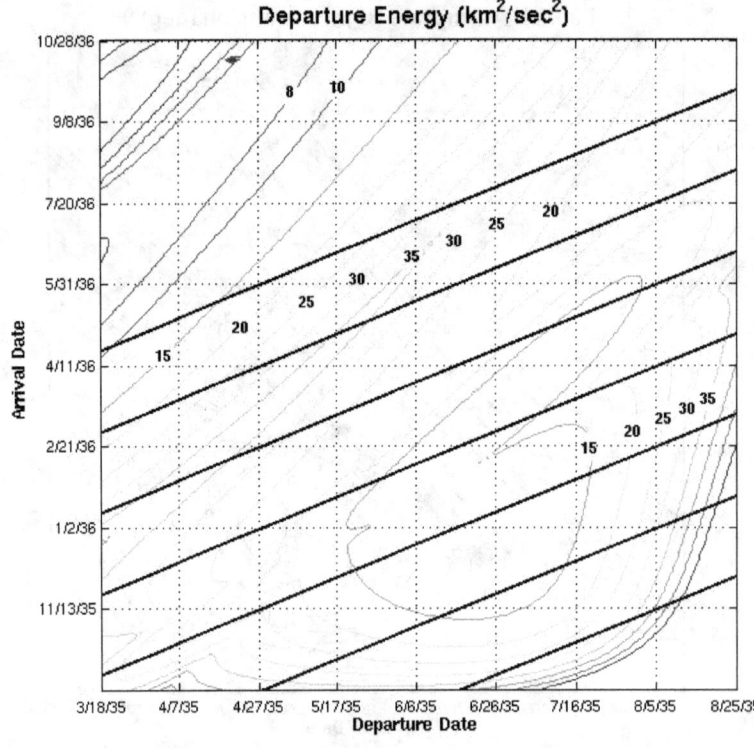

Earth-Mars Trajectories with Mid-Course Corrections
2035
Departure Energy (km^2/sec^2)

NASA/TM—2010-216764

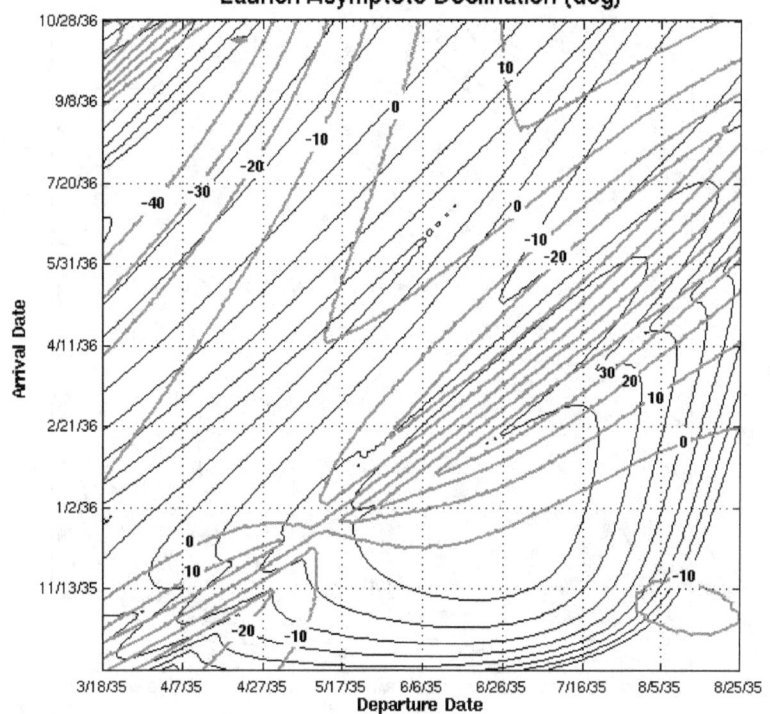
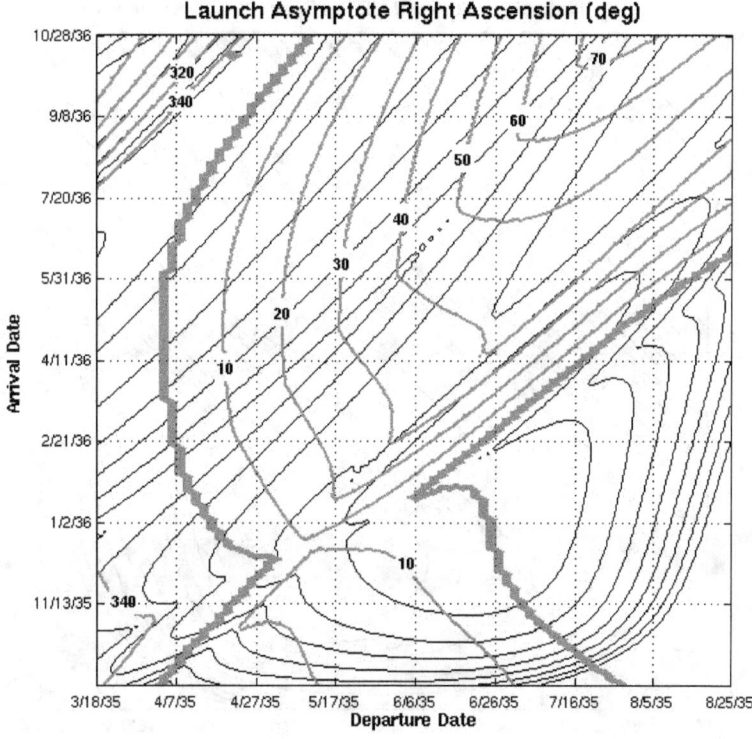

NASA/TM—2010-216764

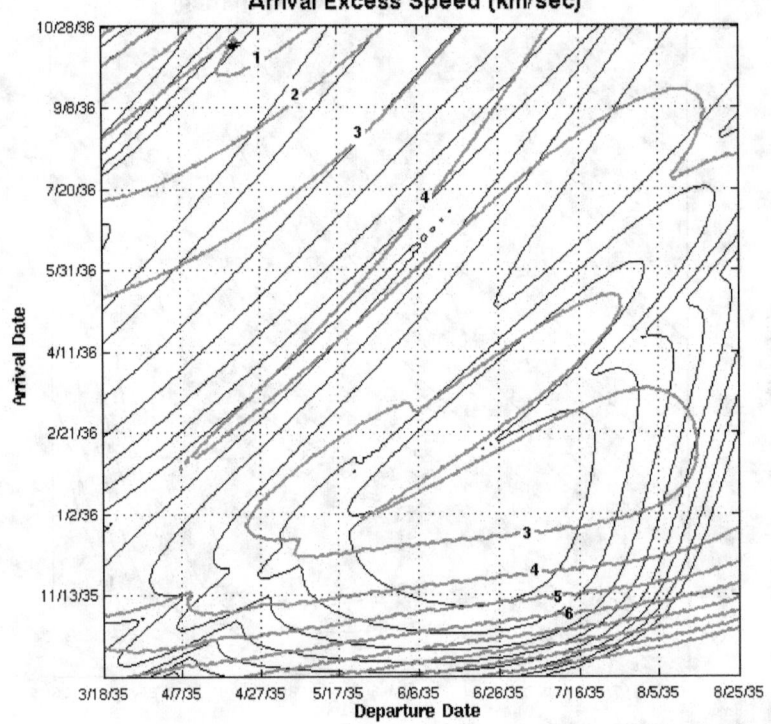

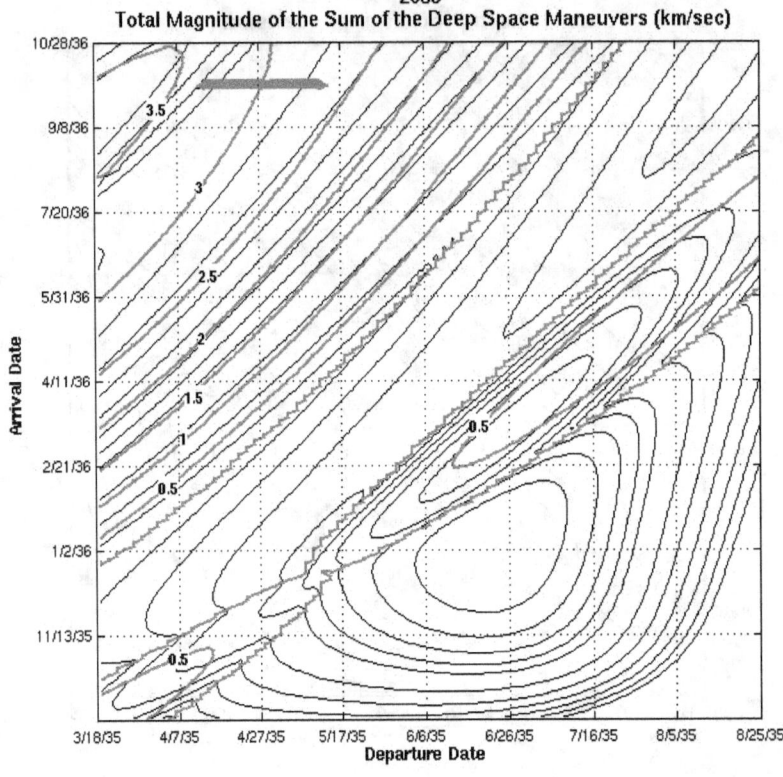

NASA/TM—2010-216764   44

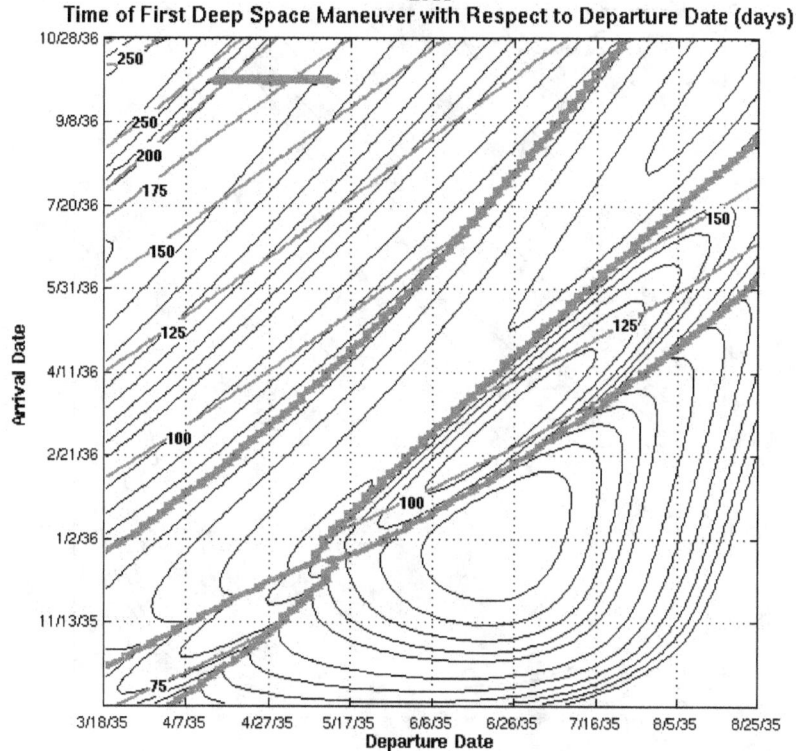

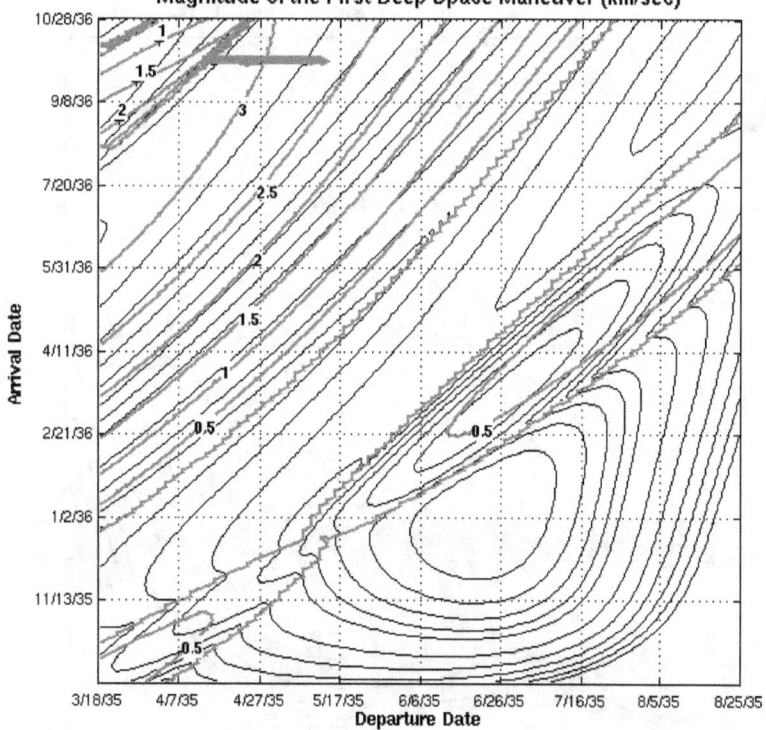

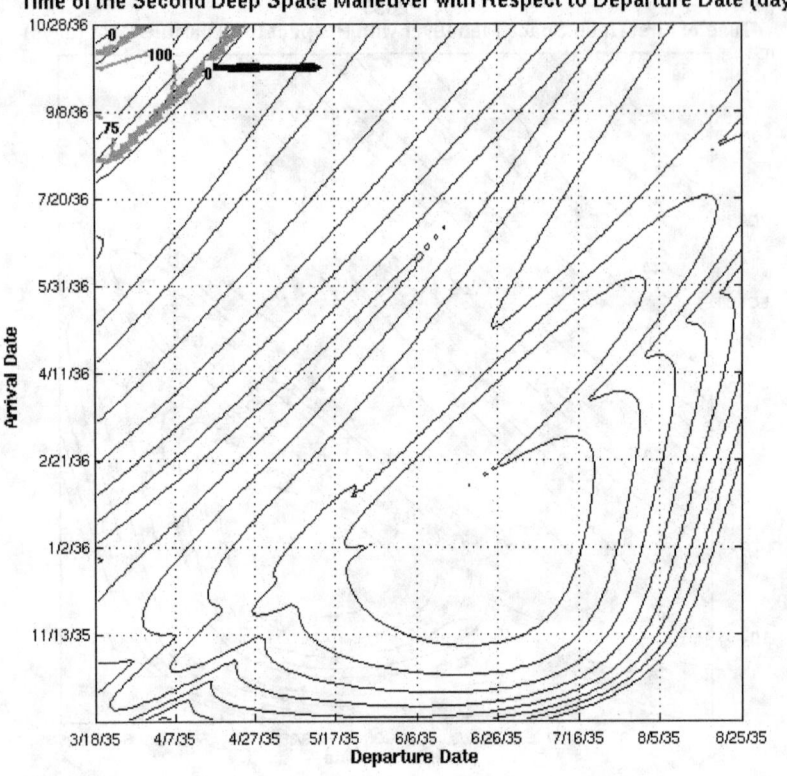

Earth-Mars Trajectories with Mid-Course Corrections
2035
Time of the Second Deep Space Maneuver with Respect to Departure Date (days)

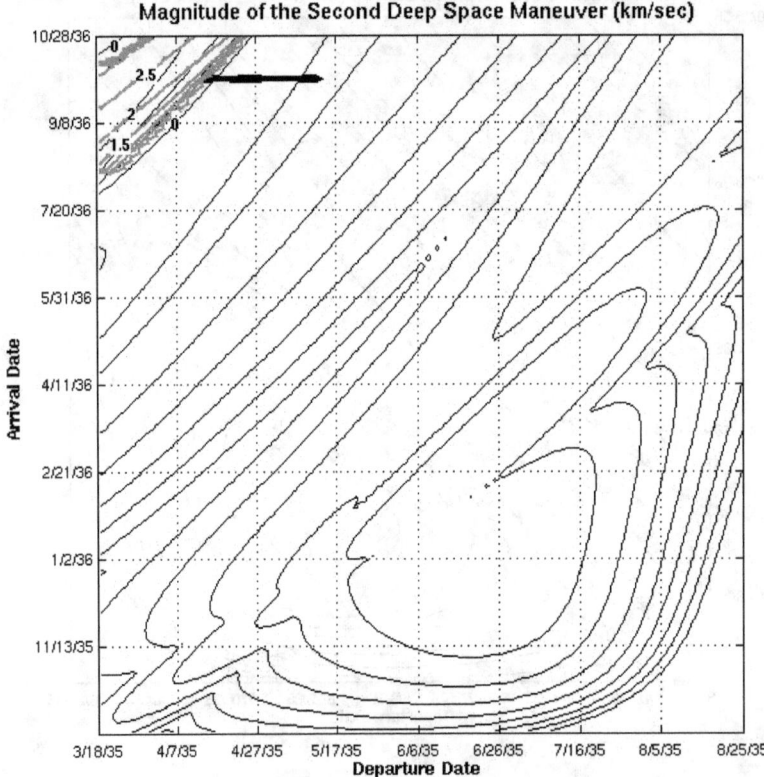

Earth-Mars Trajectories with Mid-Course Corrections
2035
Magnitude of the Second Deep Space Maneuver (km/sec)

Earth to Mars—2037 Opportunity

TABLE 7.—EARTH TO MARS—2037 OPPORTUNITY—ENERGY MINIMA

Mission type	Earth departure date (m/d/yr)	Mars arrival date (m/d/yr)	C_3 (km^2/sec^2)	Right ascension (deg)	Declination (deg)	Mars arrival excess speed (km/s)
Type 1	6/2/37	12/17/37	**17.07**	43.45	39.79	3.344
Type 2	6/18/37	7/19/38	**14.84**	74.97	13.59	3.356
Type 1	6/30/37	2/19/38	28.33	26.54	32.34	**2.334**
Type 2	4/13/37	2/7/38	31.13	66.88	1.891	**2.422**

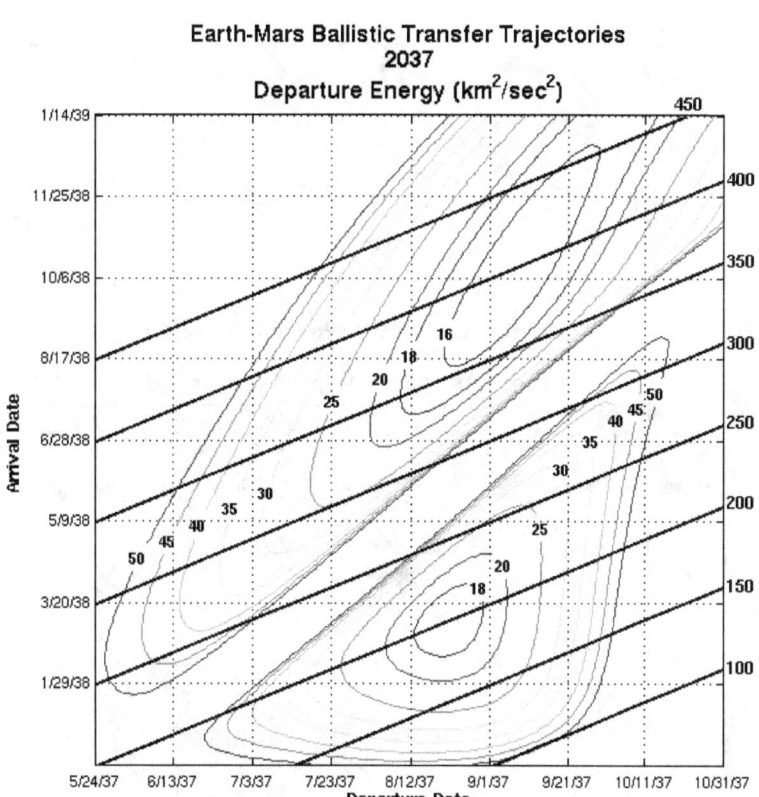

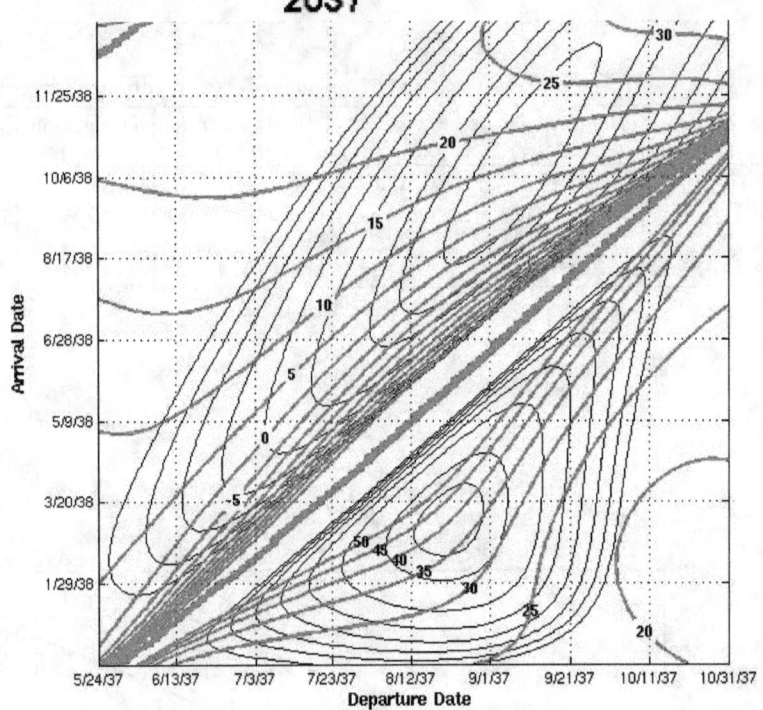

Earth-Mars Ballistic Transfer Trajectories
2037

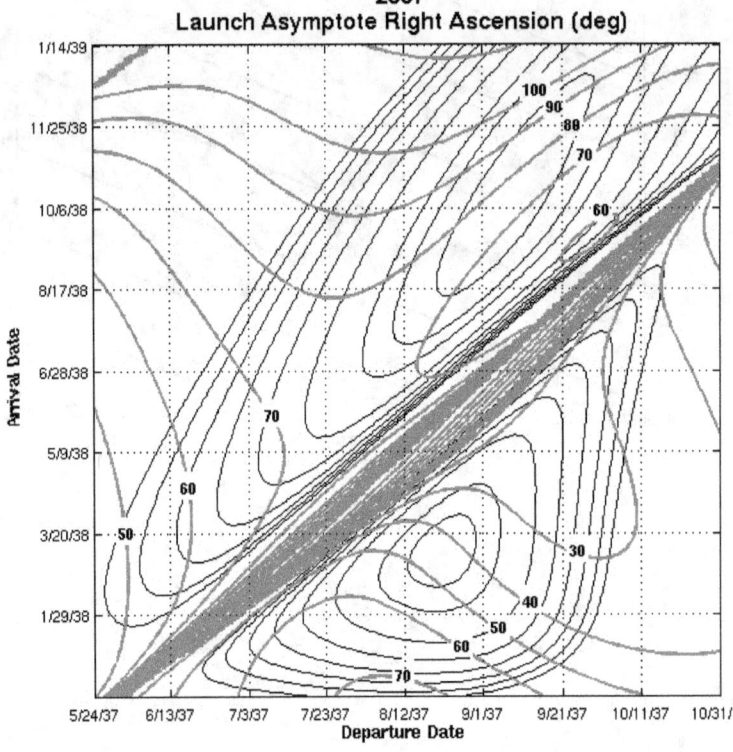

Earth-Mars Ballistic Transfer Trajectories
2037
Launch Asymptote Right Ascension (deg)

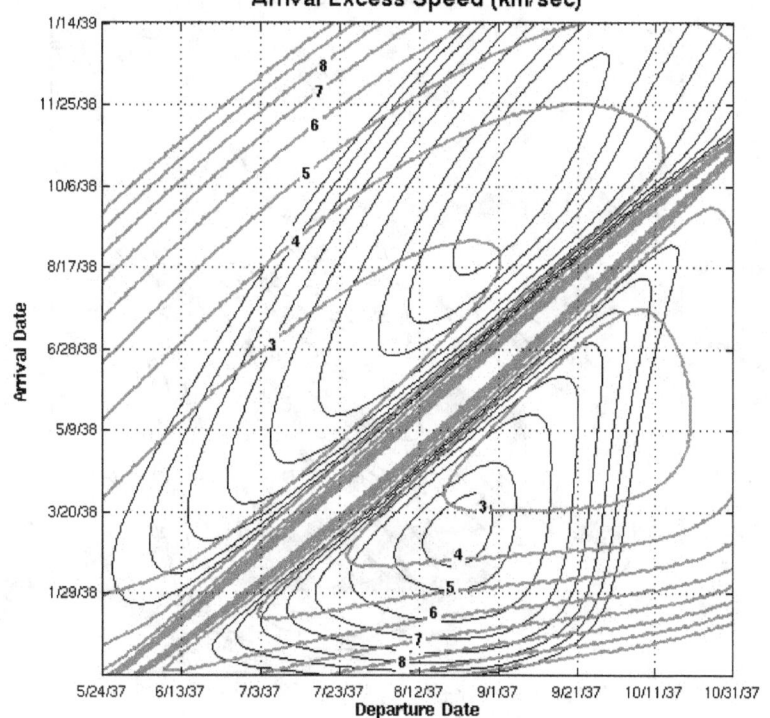

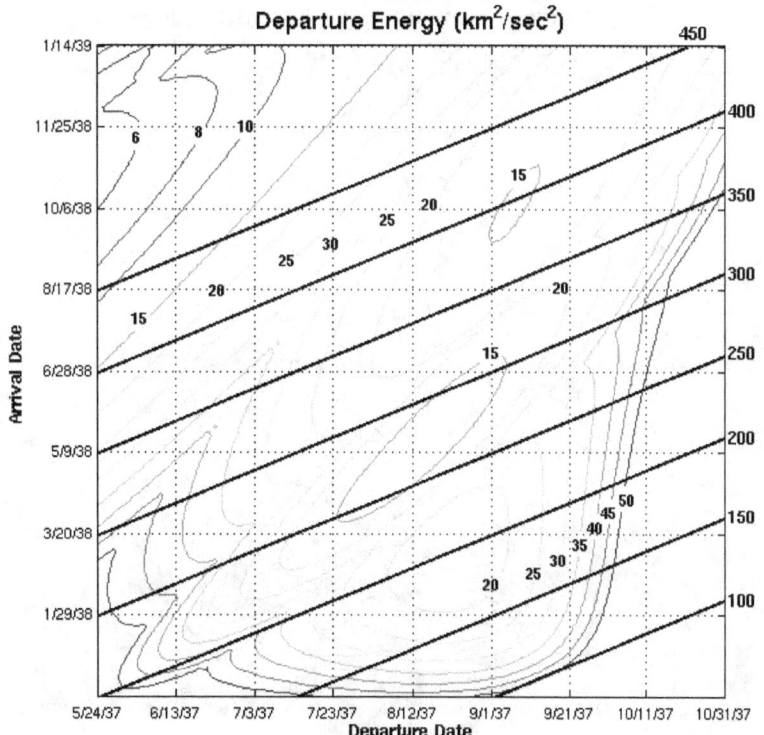

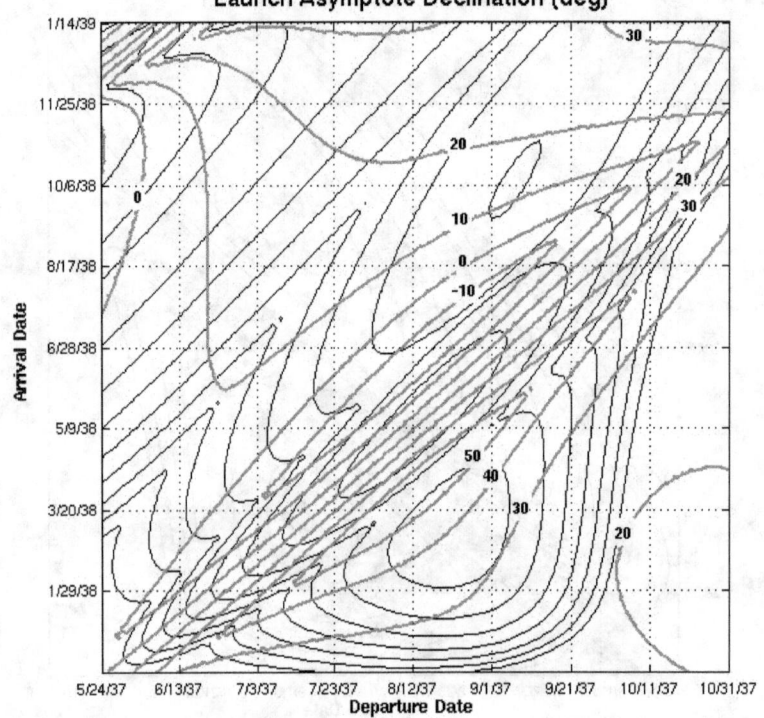

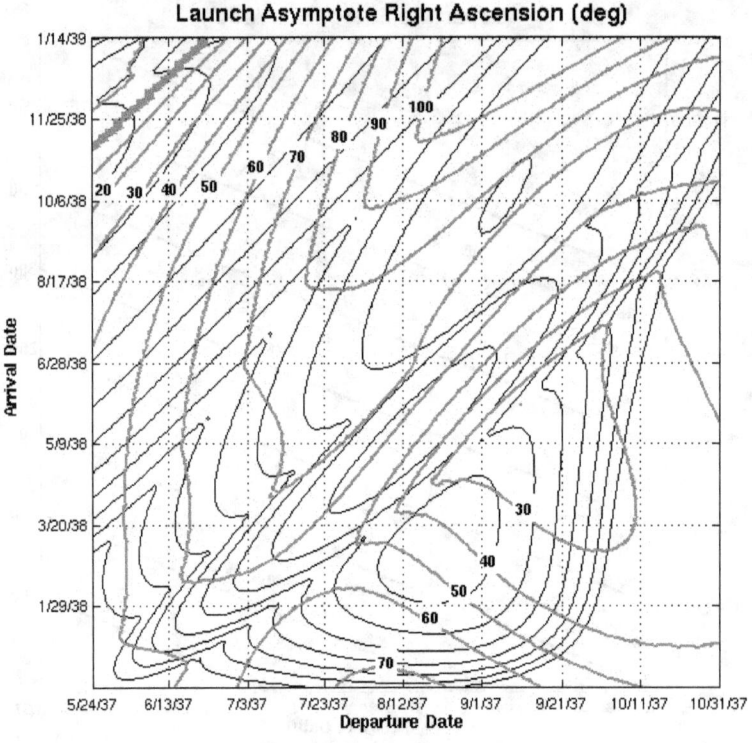

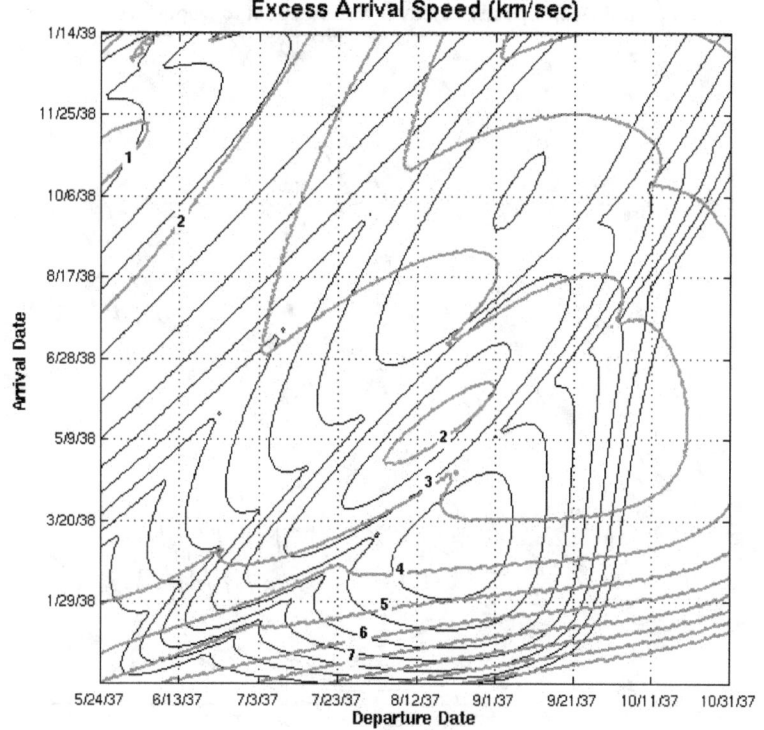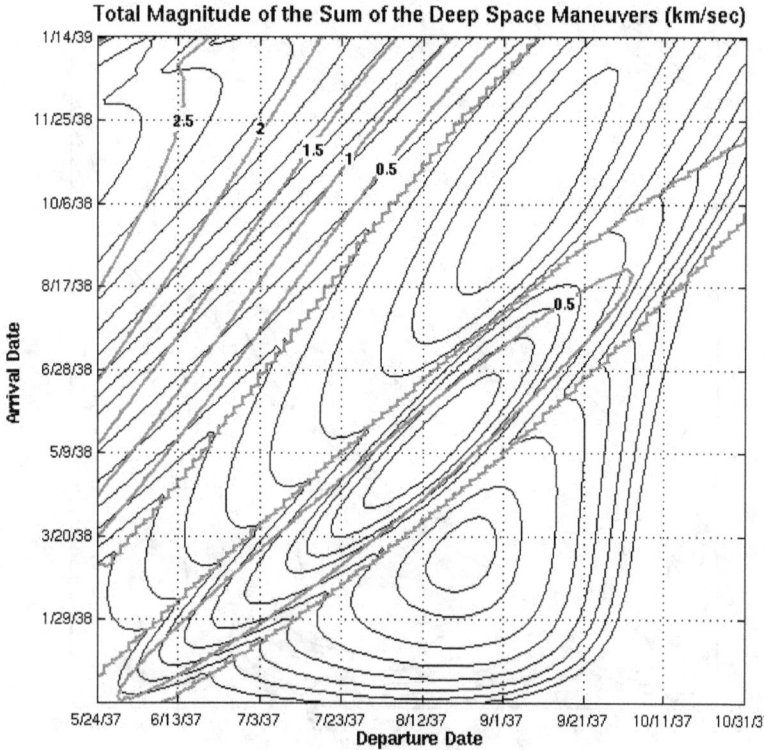

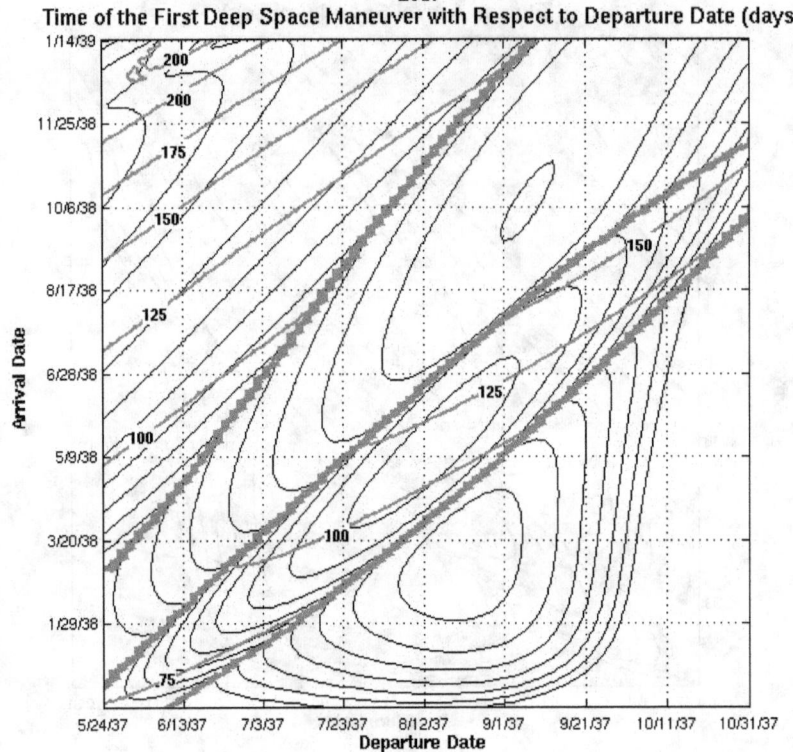

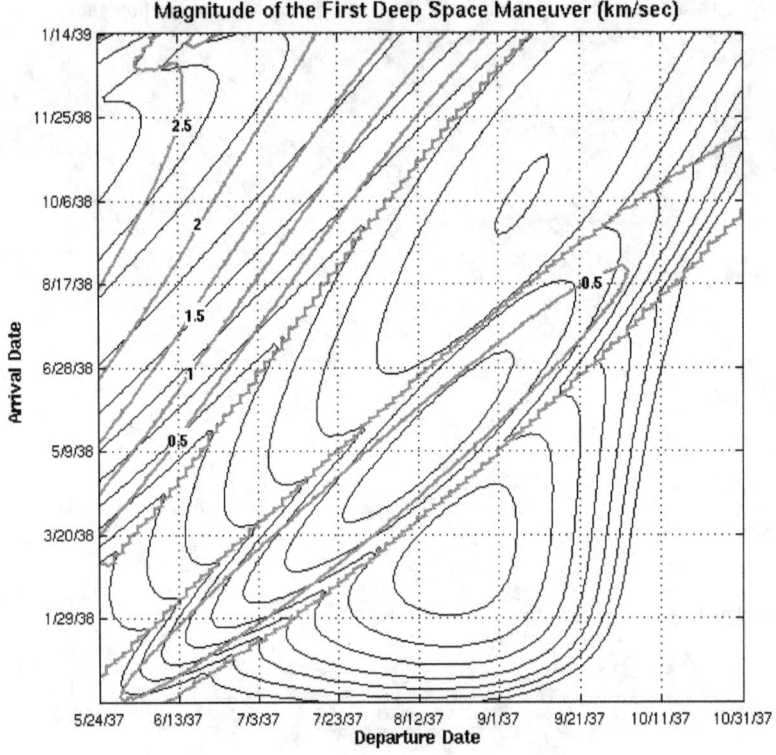

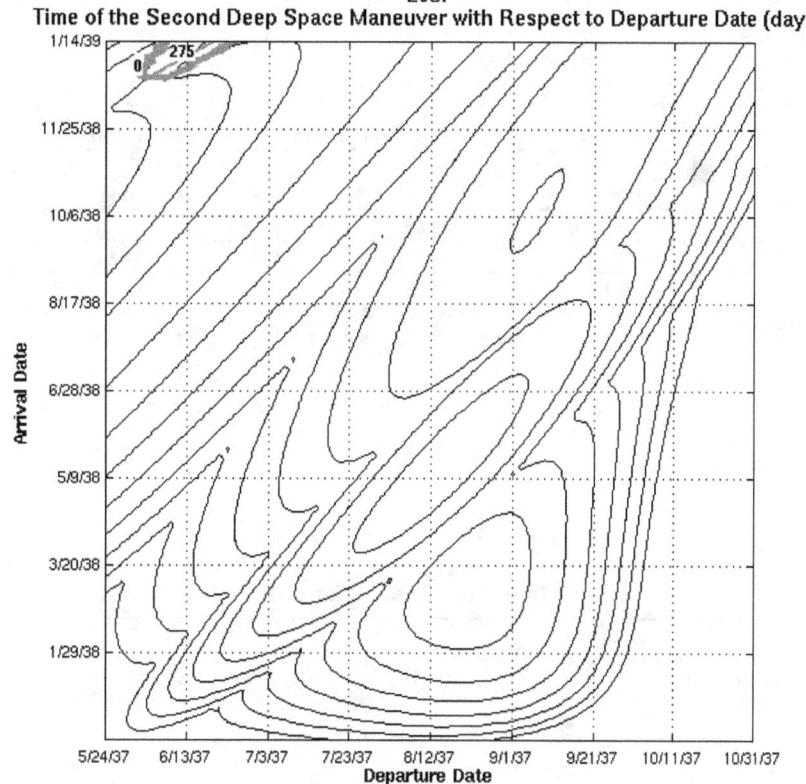

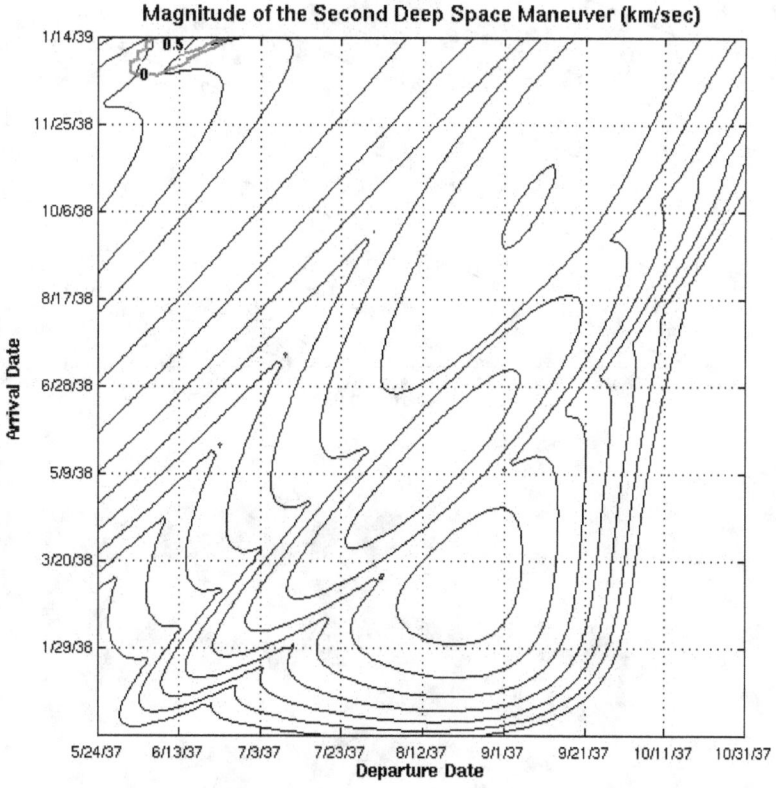

NASA/TM—2010-216764 53

Earth to Mars—2039 Opportunity

TABLE 8.—EARTH TO MARS—2039 OPPORTUNITY—ENERGY MINIMA

Mission type	Earth departure date (m/d/yr)	Mars arrival date (m/d/yr)	C_3 (km^2/sec^2)	Right ascension (deg)	Declination (deg)	Mars arrival excess speed (km/s)
Type 1	7/19/39	2/16/40	**18.65**	82.89	49.48	4.03
Type 2	7/15/39	7/9/40	**12.17**	94.37	18.66	2.701
Type 1	8/12/39	5/4/40	29.07	63.6	40.89	**2.361**
Type 2	6/11/39	5/10/40	19.29	101.7	10.12	**2.384**

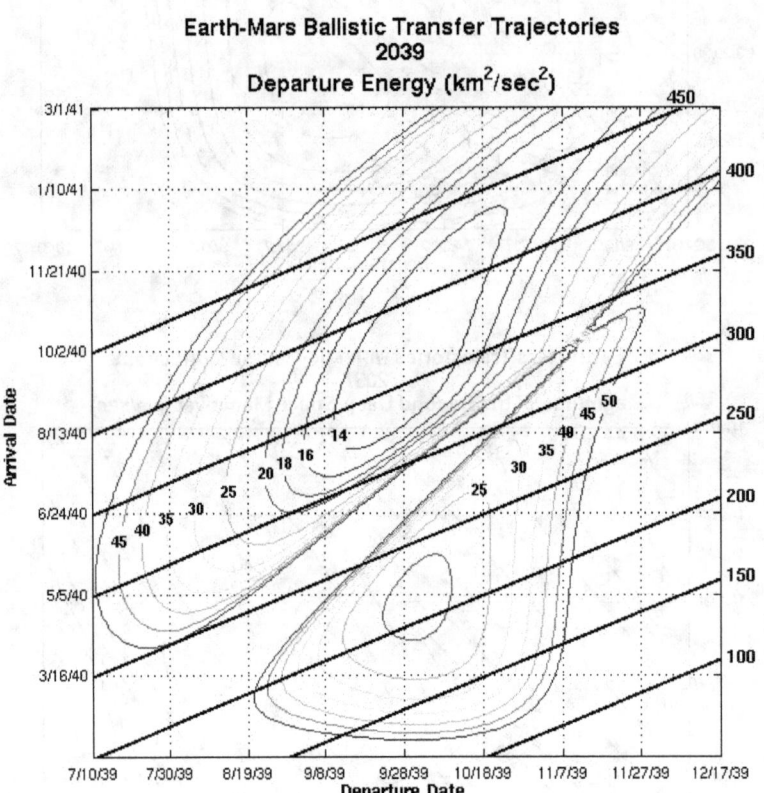

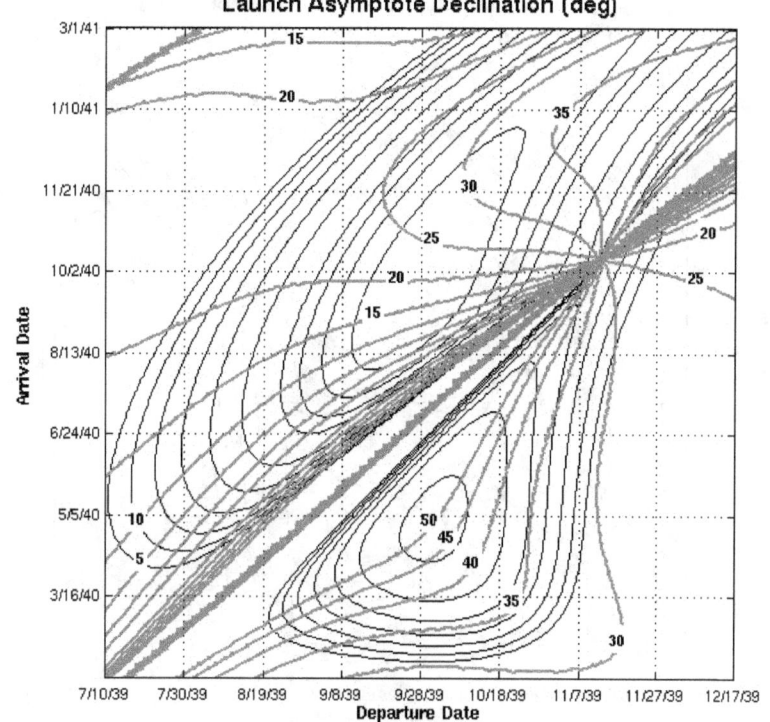

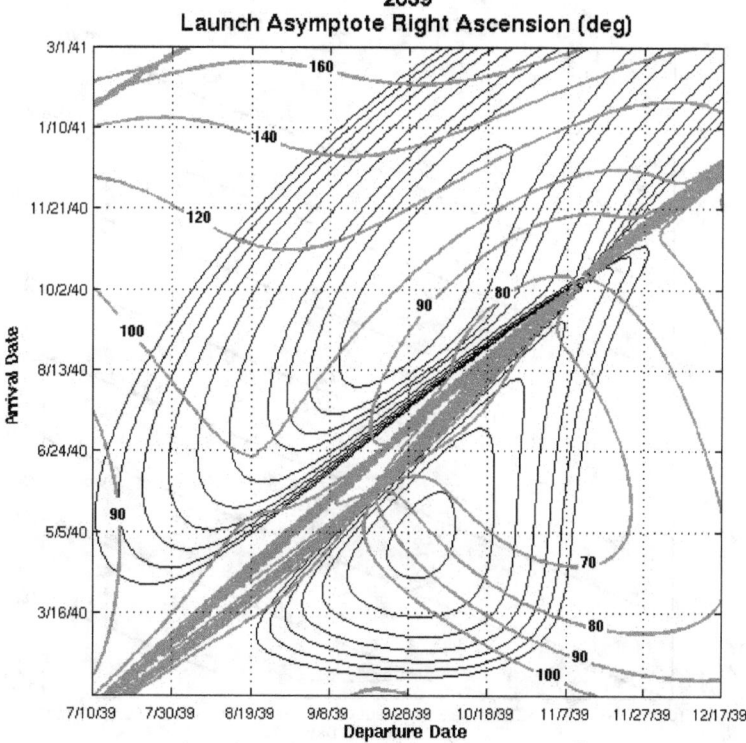

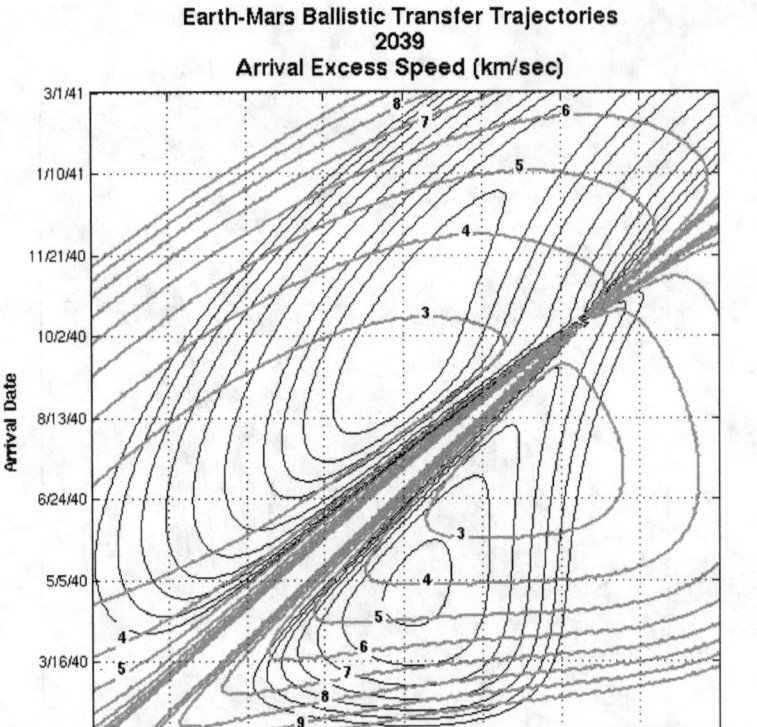
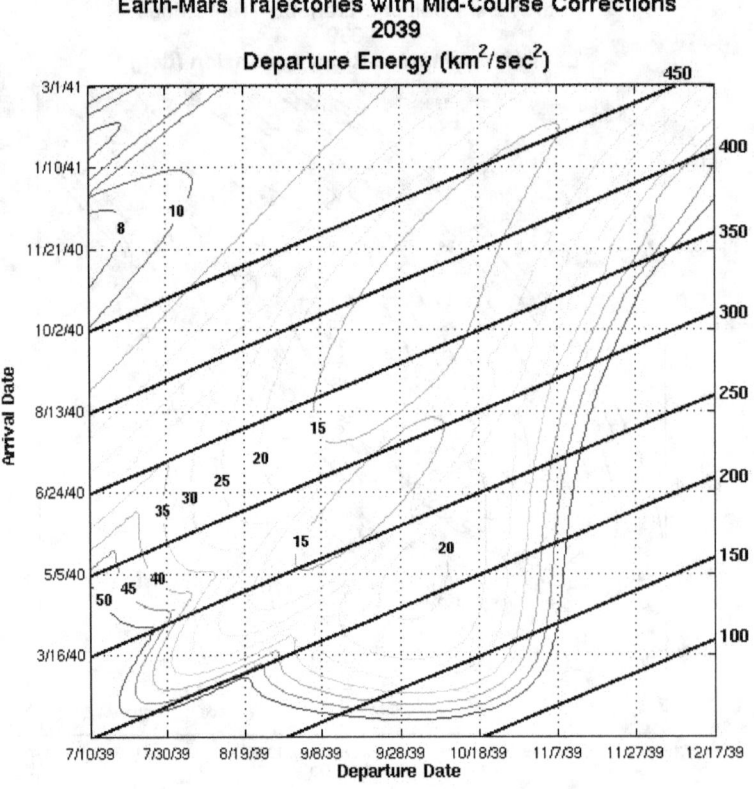

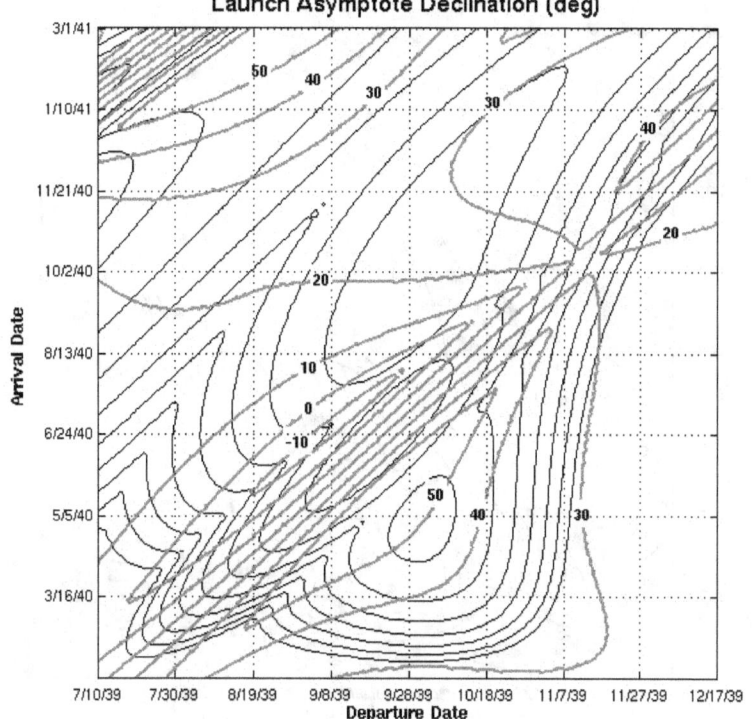

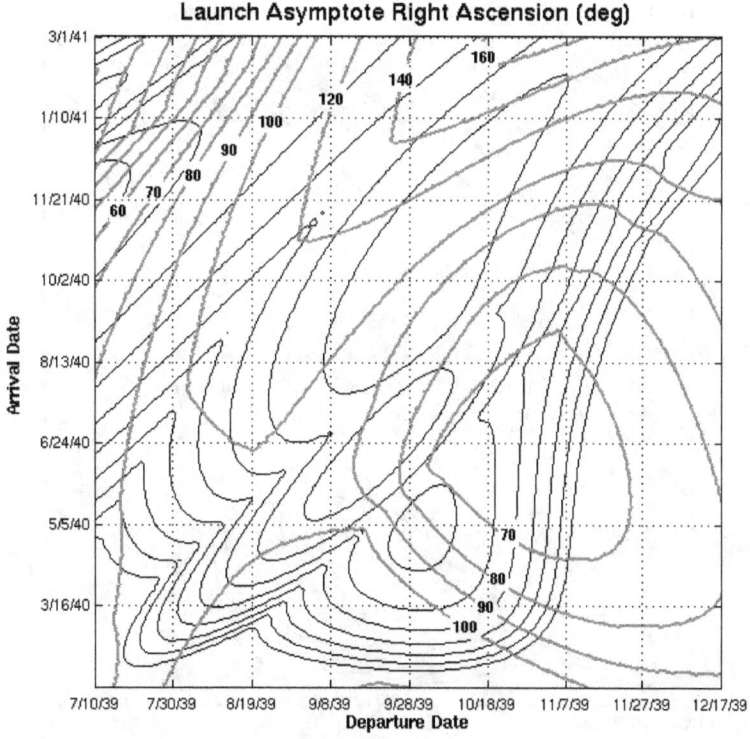

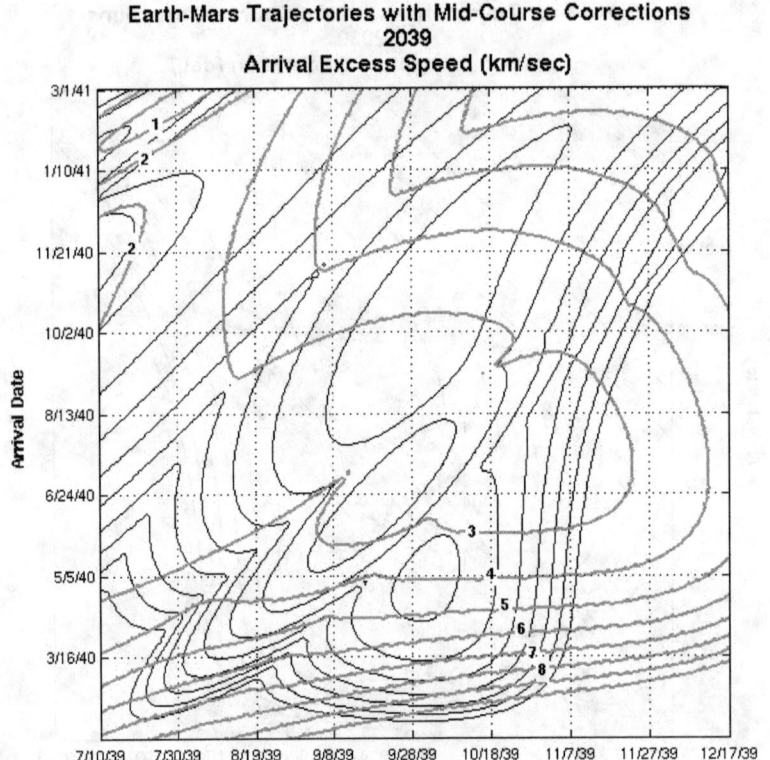

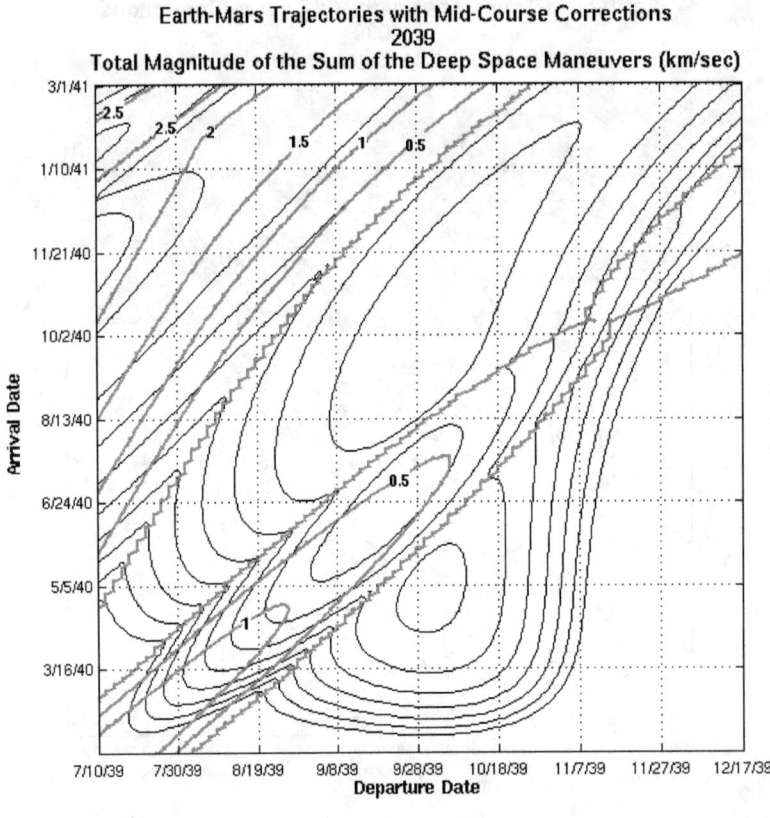

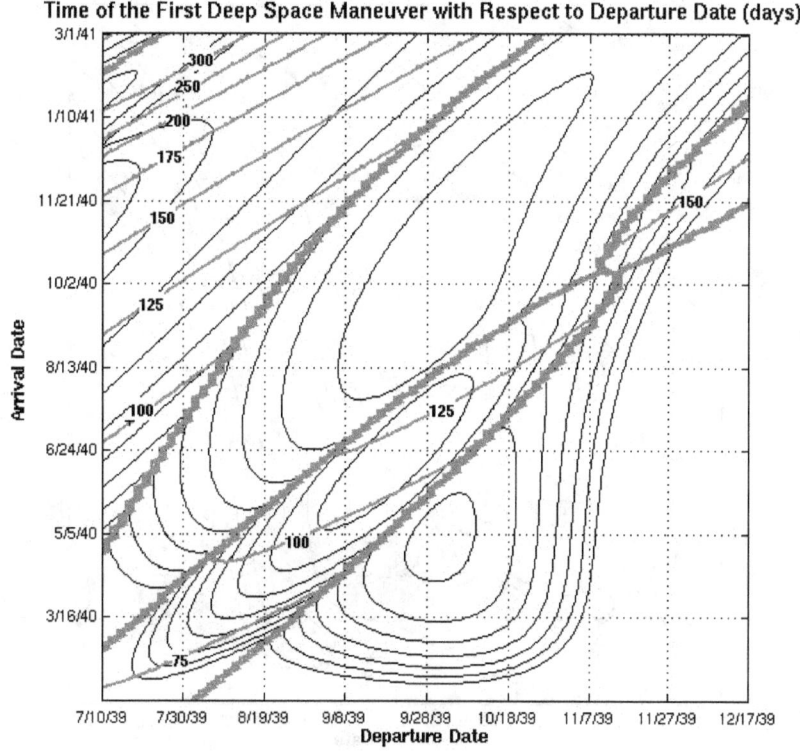

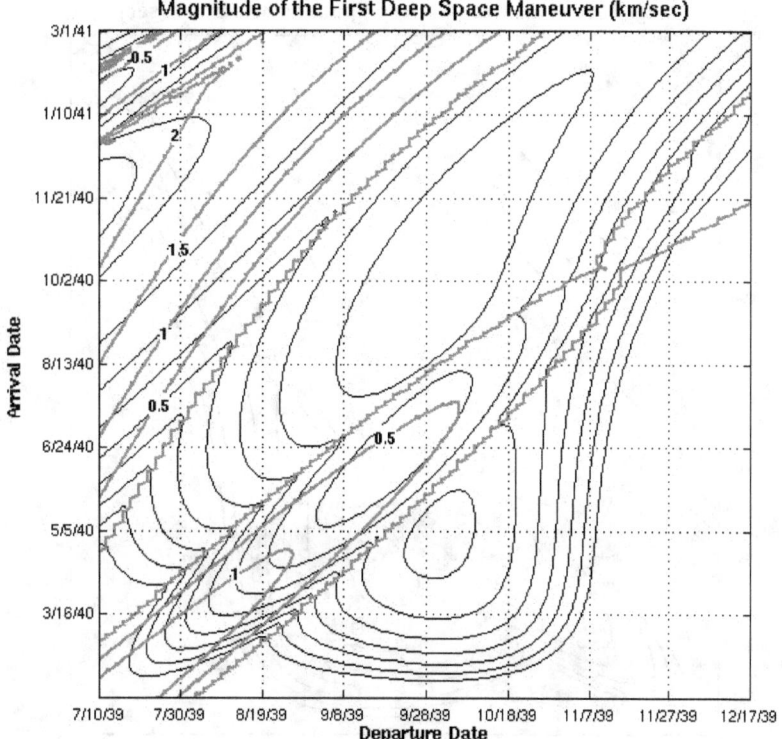

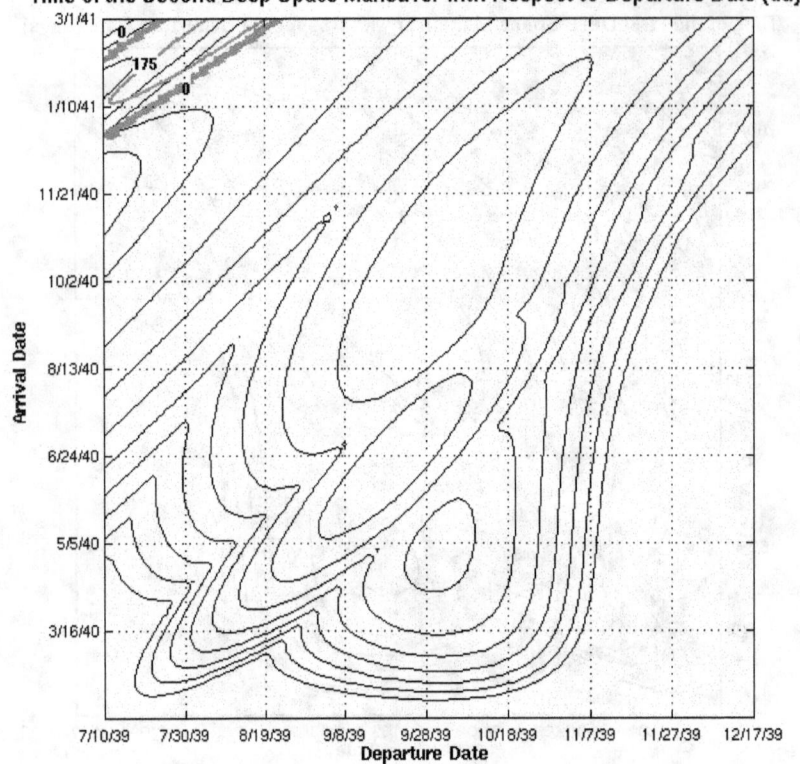

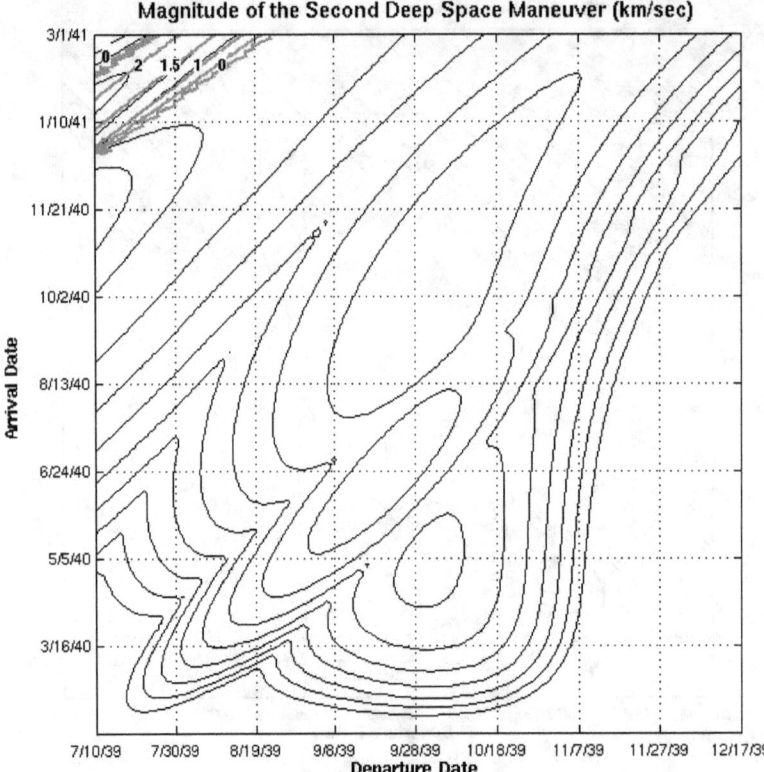

Earth to Mars—2041 Opportunity

TABLE 9.—EARTH TO MARS—2041 OPPORTUNITY—ENERGY MINIMA

Mission type	Earth departure date (m/d/yr)	Mars arrival date (m/d/yr)	C_3 (km^2/sec^2)	Right ascension (deg)	Declination (deg)	Mars arrival excess speed (km/s)
Type 1	10/31/41	6/20/42	**14.86**	119.6	46.18	4.016
Type 2	10/19/41	8/31/42	**9.818**	119.9	21.71	2.49
Type 1	11/12/41	8/31/42	15.84	104.2	39.59	**2.489**
Type 2	10/21/41	9/4/42	9.819	119.1	23.5	**2.483**

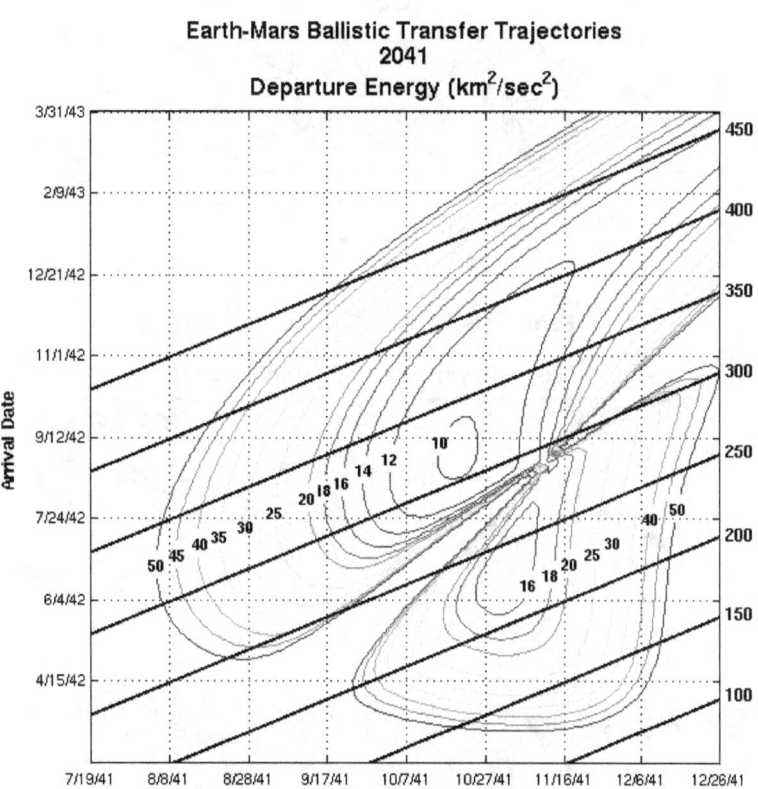

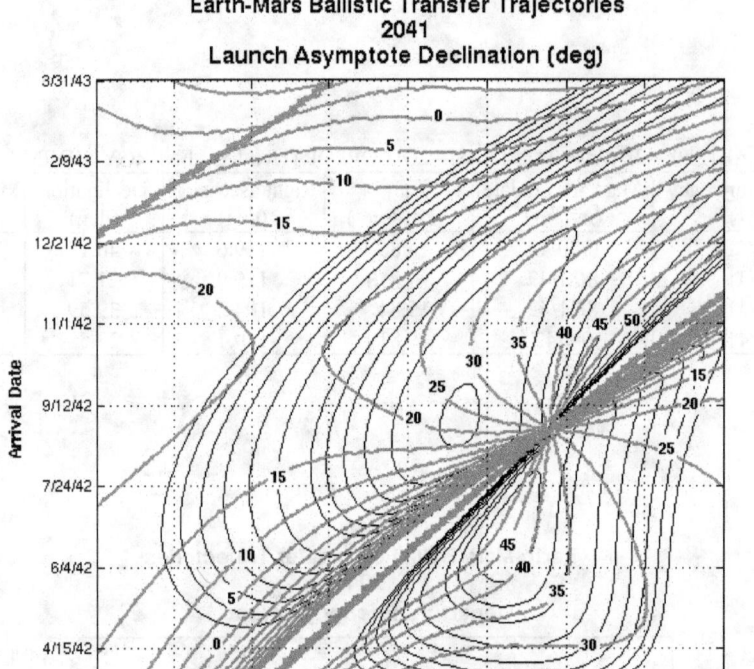

Earth-Mars Ballistic Transfer Trajectories
2041
Launch Asymptote Declination (deg)

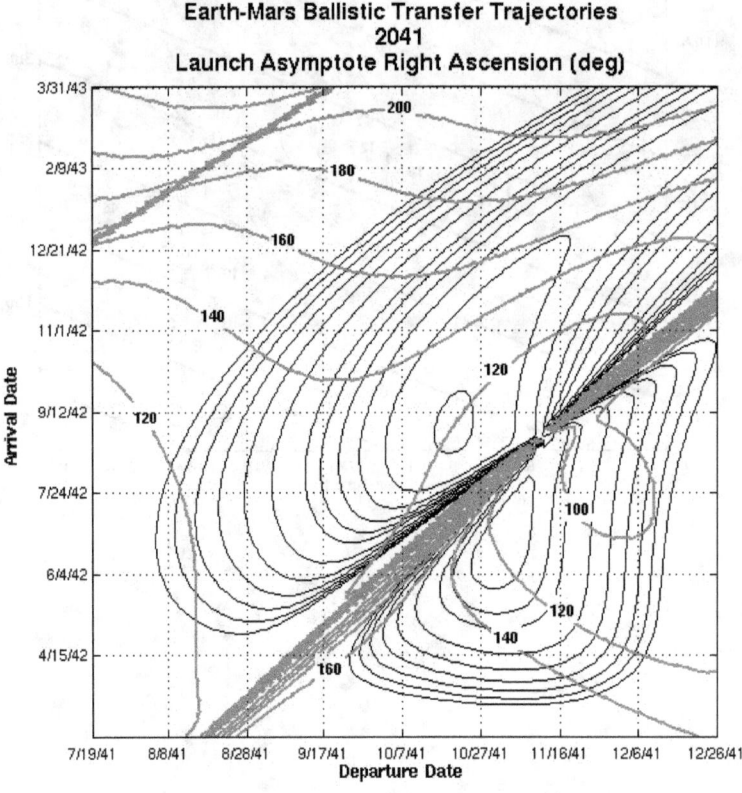

Earth-Mars Ballistic Transfer Trajectories
2041
Launch Asymptote Right Ascension (deg)

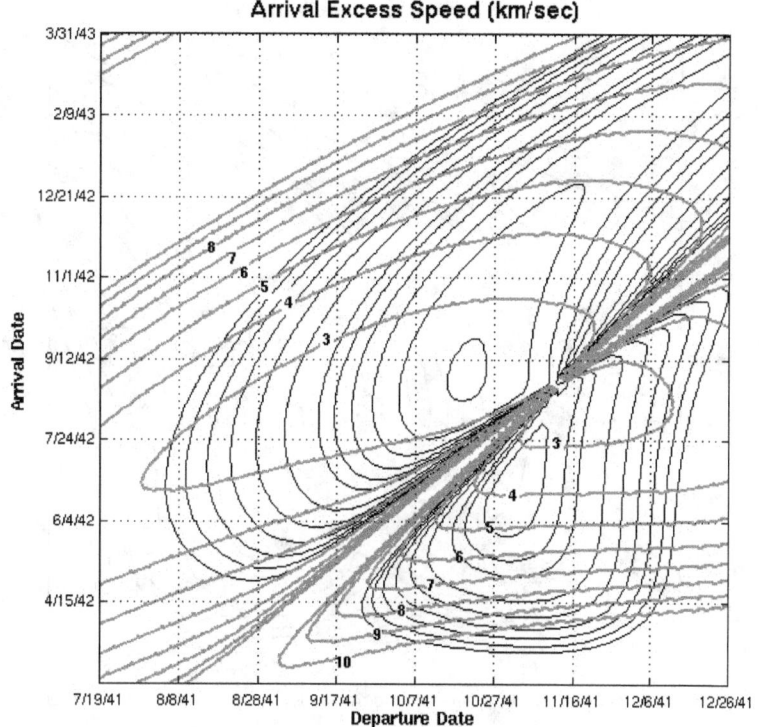

Earth-Mars Ballistic Transfer Trajectories
2041
Arrival Excess Speed (km/sec)

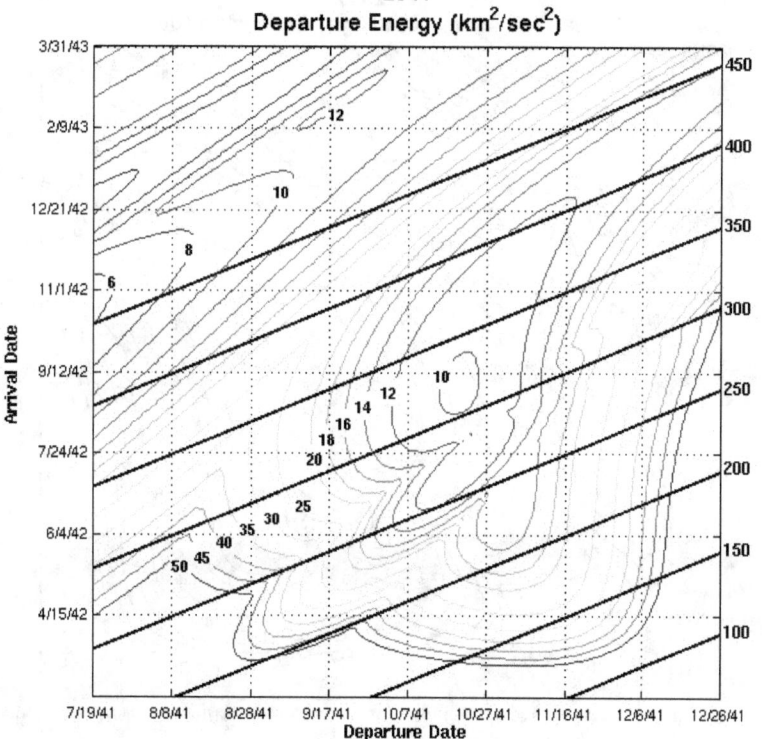

Earth-Mars Trajectories with Mid-Course Corrections
2041
Departure Energy (km^2/sec^2)

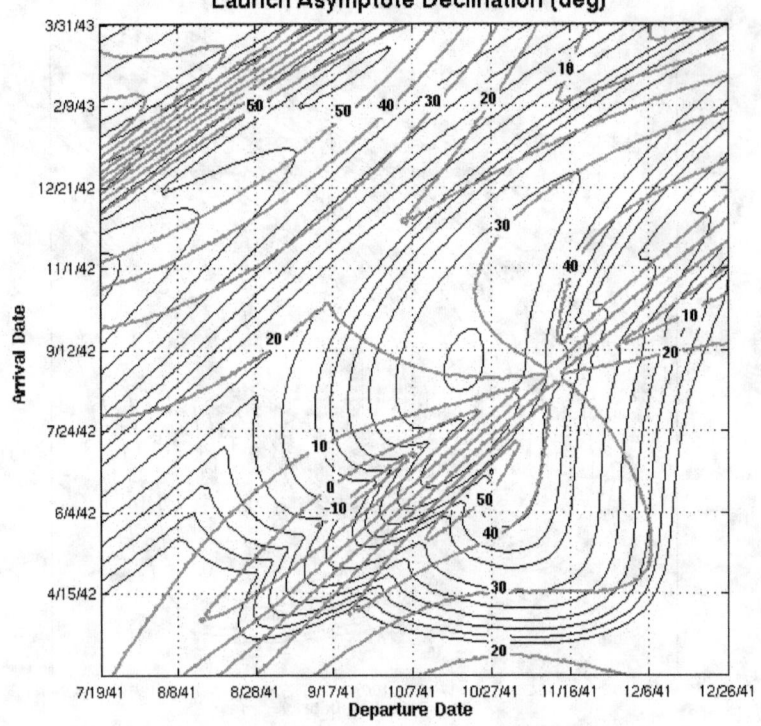

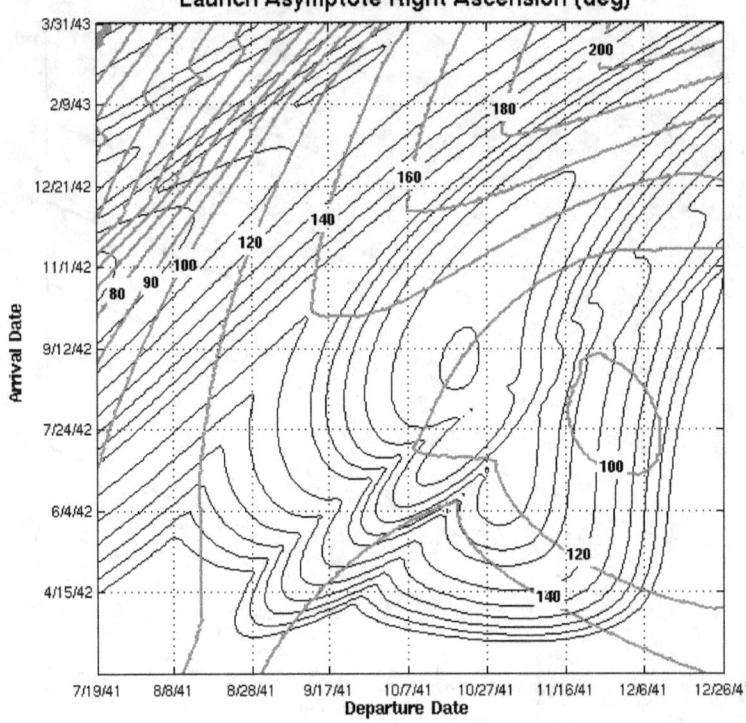

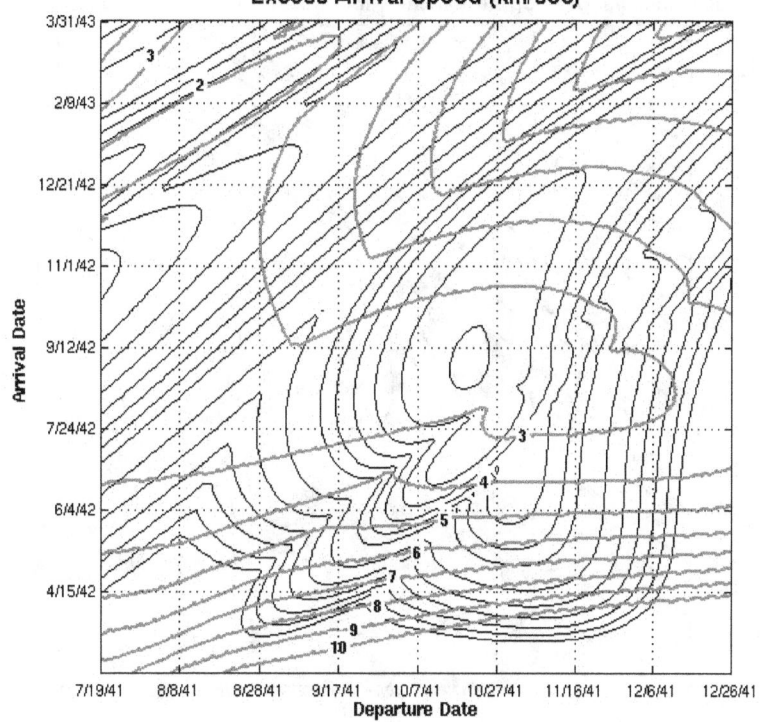

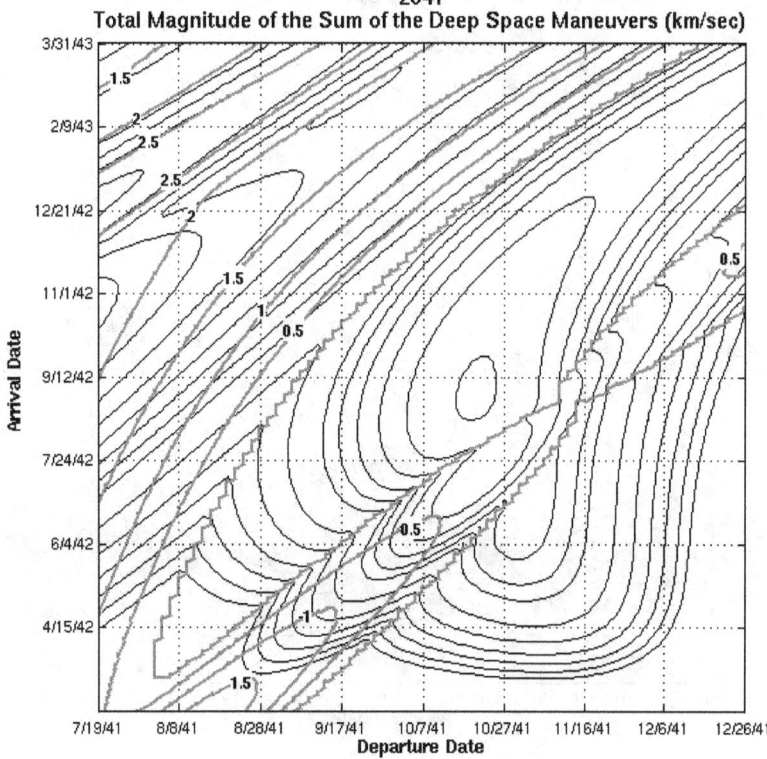

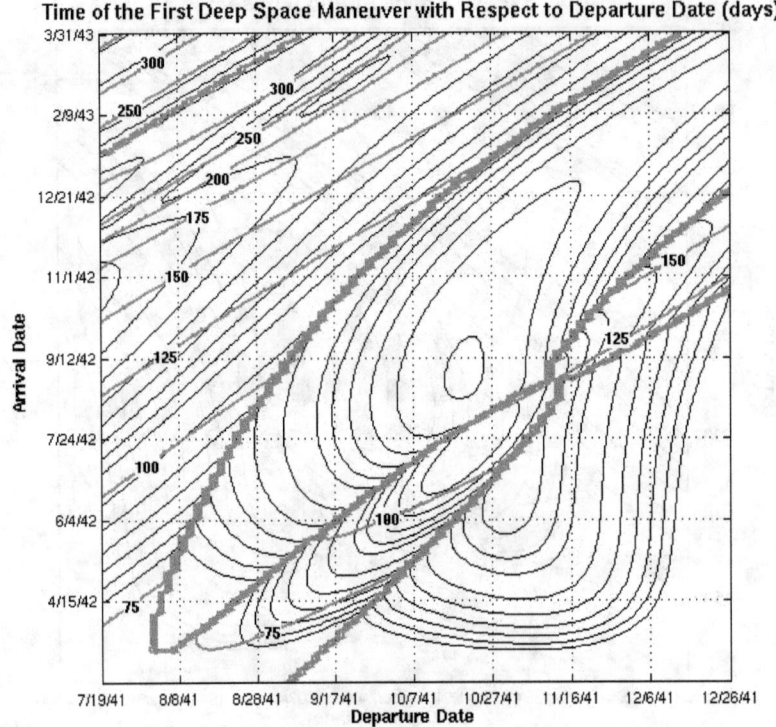
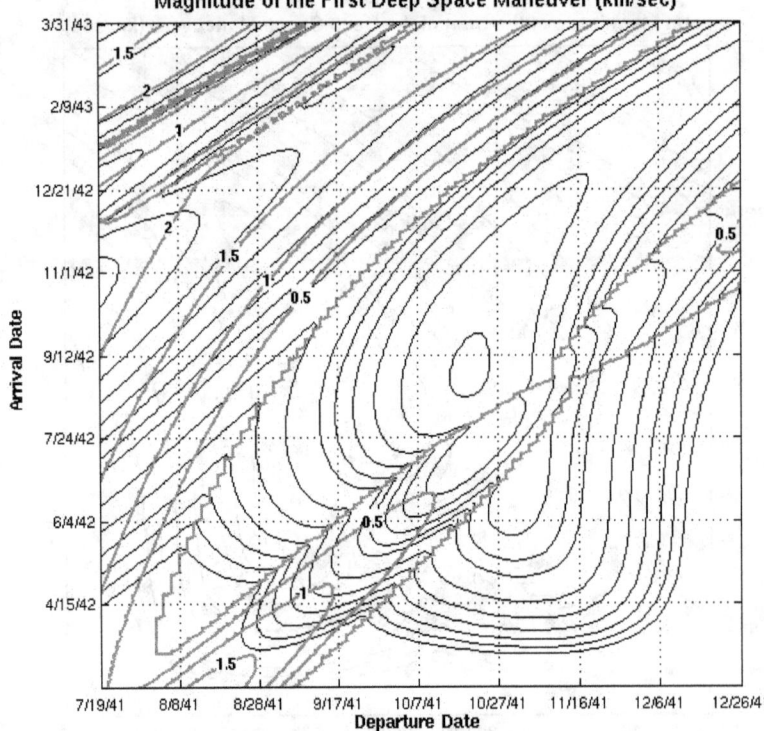

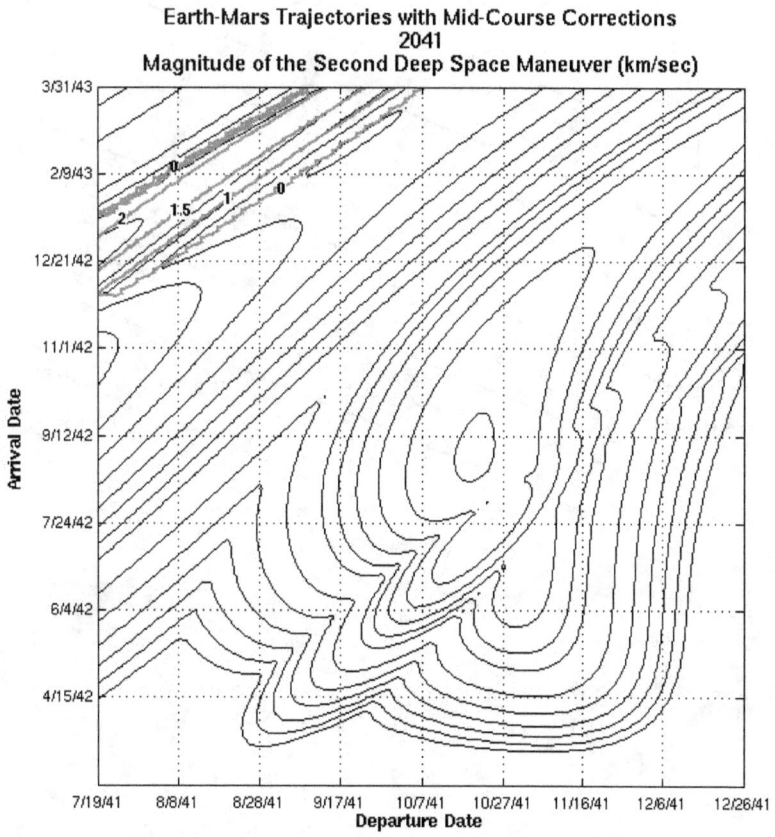

Earth to Mars—2043 Opportunity

TABLE 10.—EARTH TO MARS—2043 OPPORTUNITY—ENERGY MINIMA

Mission type	Earth departure date (m/d/yr)	Mars arrival date (m/d/yr)	C_3 (km^2/sec^2)	Right ascension (deg)	Declination (deg)	Mars arrival excess speed (km/s)
Type 1	11/23/43	7/20/44	**9.032**	149.10	6.097	4.263
Type 2	11/15/43	9/16/44	**8.969**	163.80	31.18	2.793
Type 1	12/31/43	8/29/44	22.88	163.80	31.18	**3.386**
Type 2	11/13/43	9/14/44	9.006	164.30	29.53	**2.793**

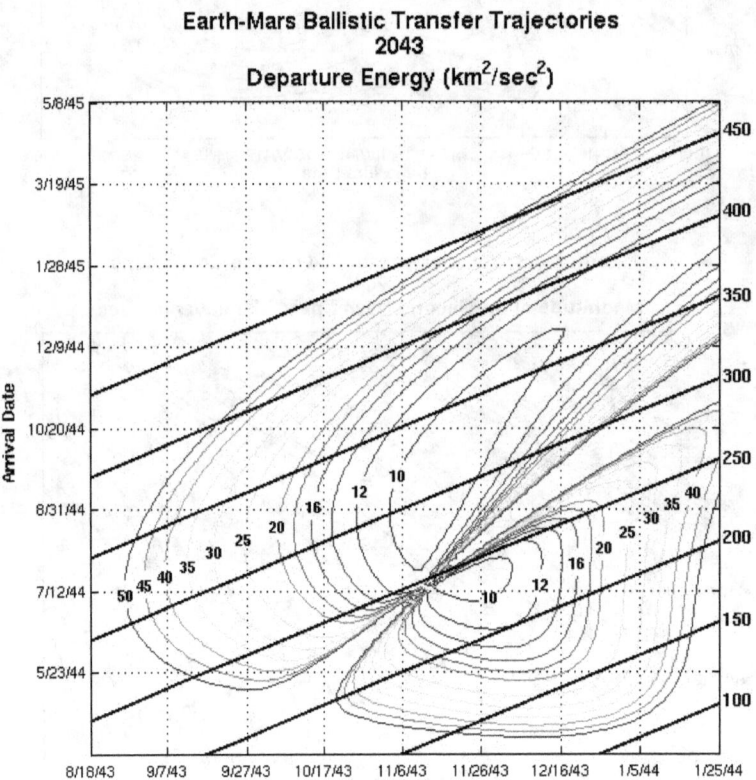

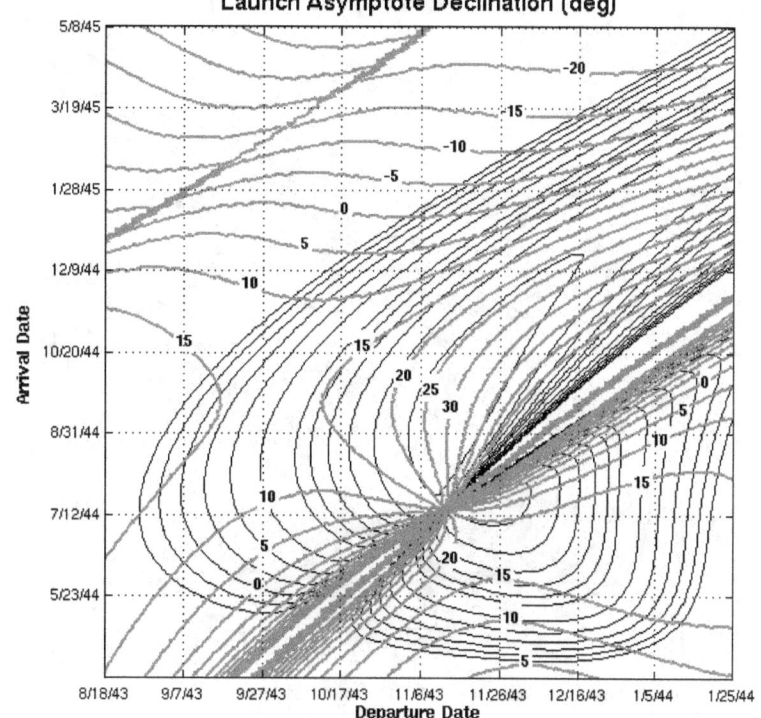

Earth-Mars Ballistic Transfer Trajectories
2043
Launch Asymptote Declination (deg)

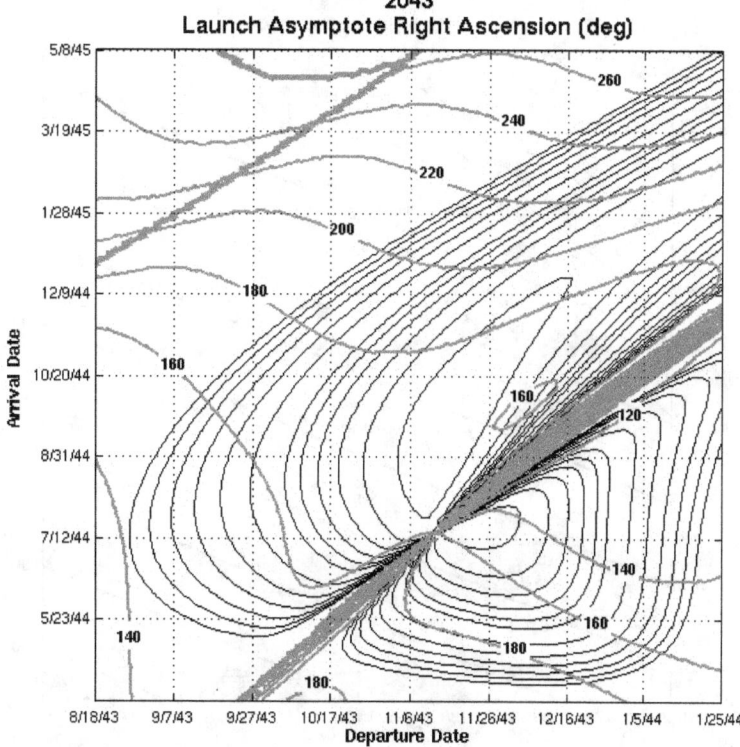

Earth-Mars Ballistic Transfer Trajectories
2043
Launch Asymptote Right Ascension (deg)

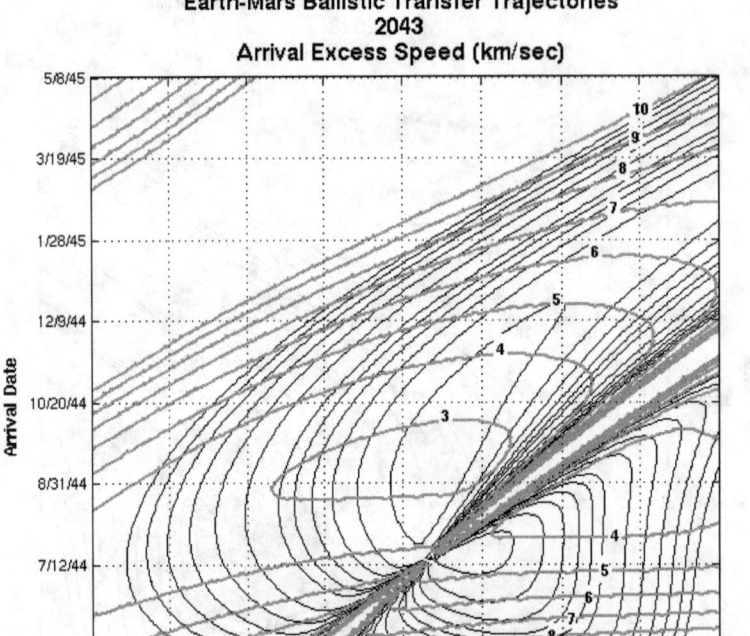

Earth-Mars Ballistic Transfer Trajectories
2043
Arrival Excess Speed (km/sec)

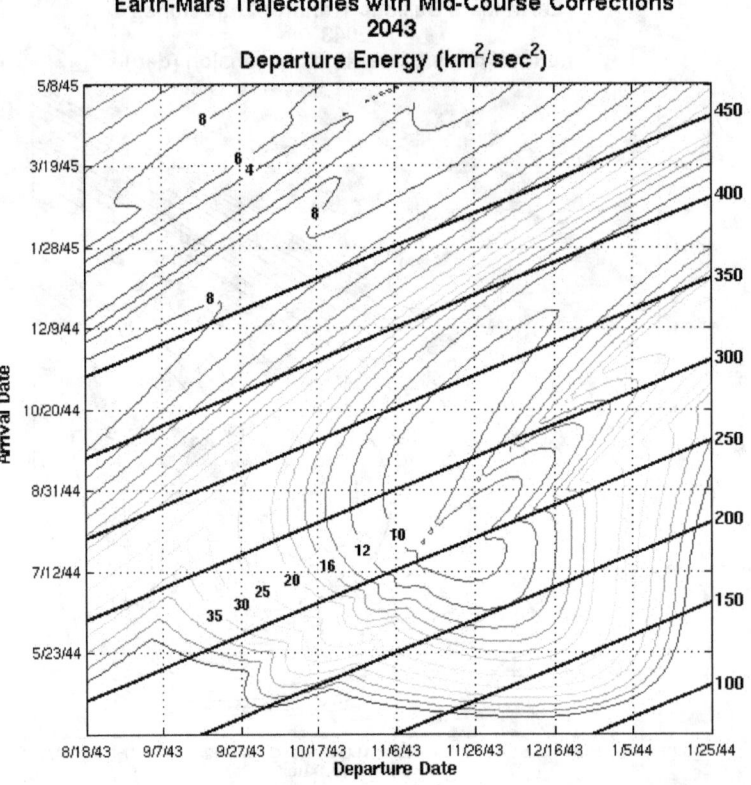

Earth-Mars Trajectories with Mid-Course Corrections
2043
Departure Energy (km²/sec²)

NASA/TM—2010-216764

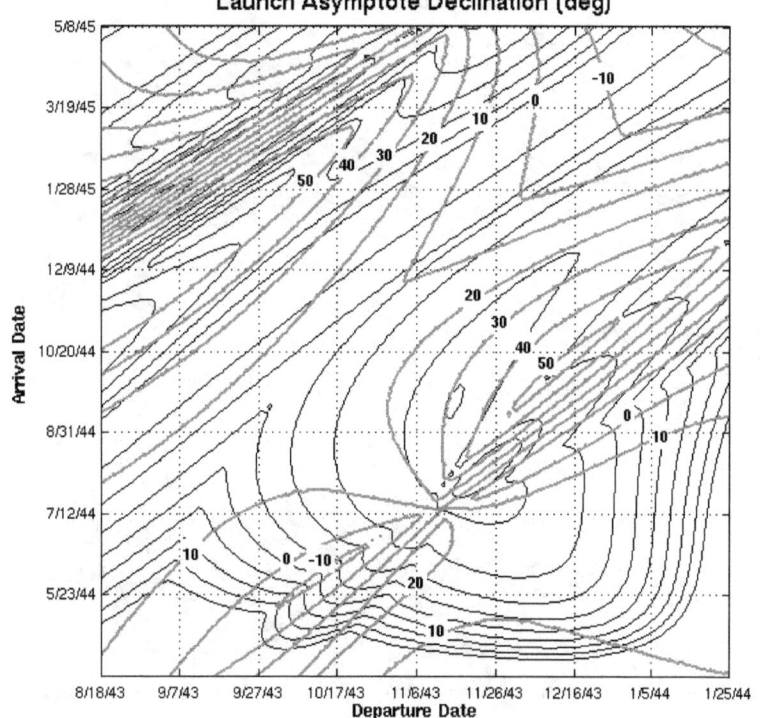

Earth-Mars Trajectories with Mid-Course Corrections
2043
Launch Asymptote Declination (deg)

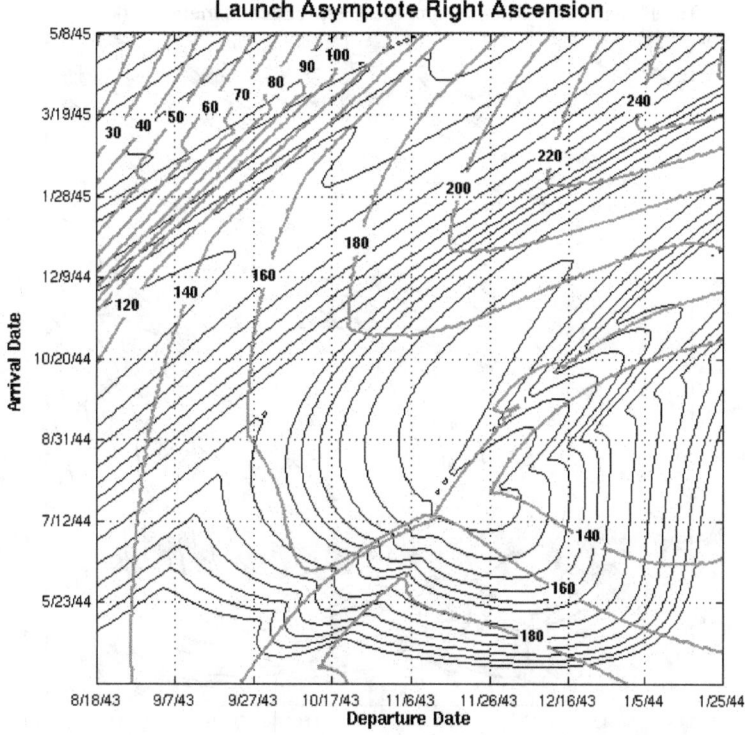

Earth-Mars Trajectories with Mid-Course Corrections
2043
Launch Asymptote Right Ascension

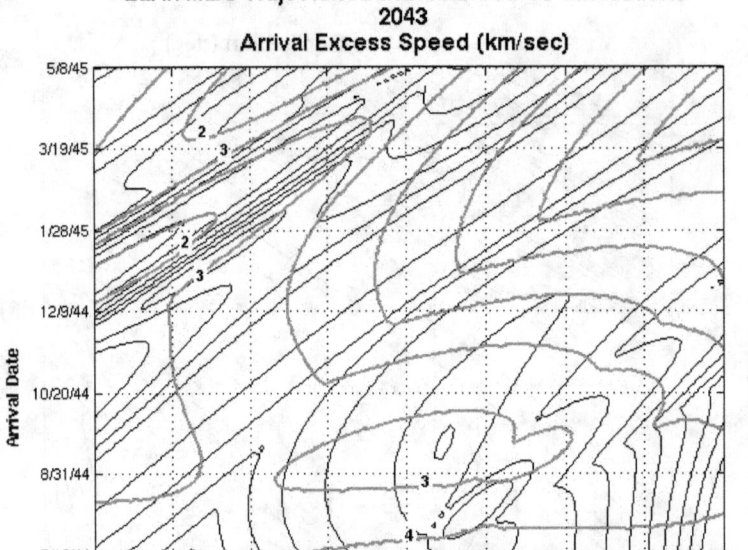

Earth-Mars Trajectories with Mid-Course Corrections
2043
Arrival Excess Speed (km/sec)

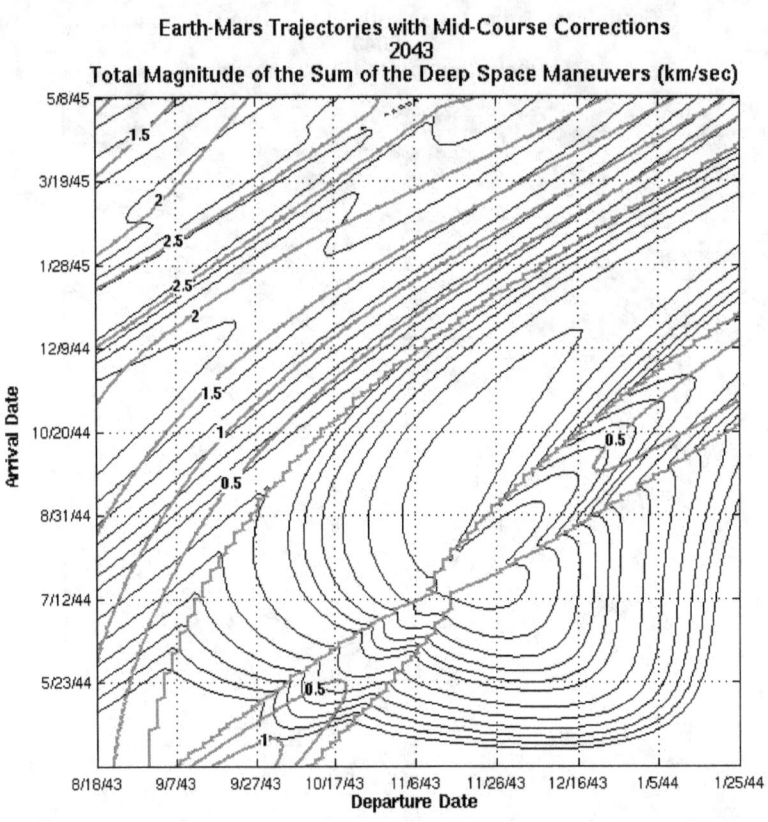

Earth-Mars Trajectories with Mid-Course Corrections
2043
Total Magnitude of the Sum of the Deep Space Maneuvers (km/sec)

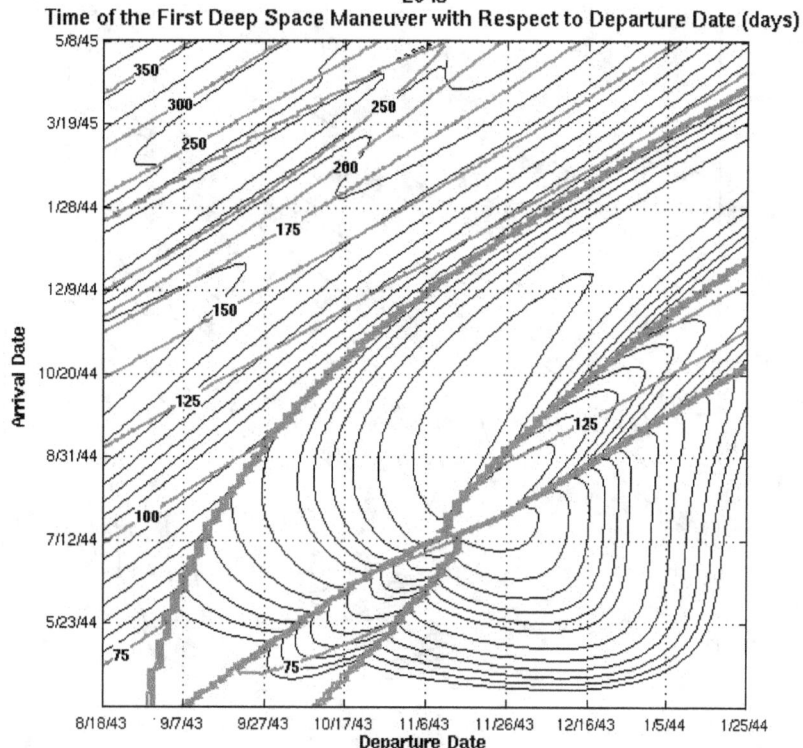

Earth-Mars Trajectories with Mid-Course Corrections
2043
Time of the First Deep Space Maneuver with Respect to Departure Date (days)

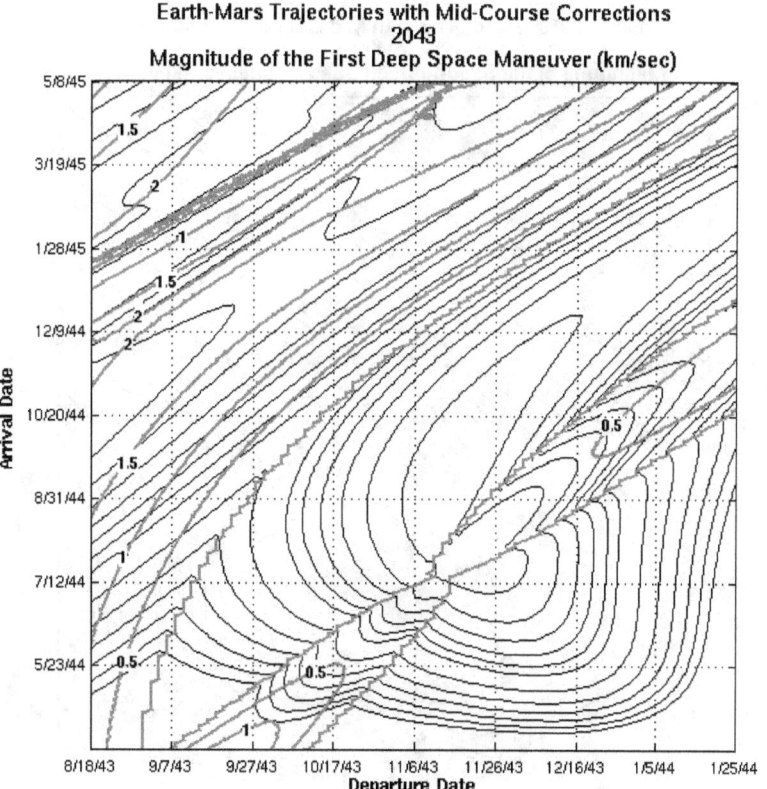

Earth-Mars Trajectories with Mid-Course Corrections
2043
Magnitude of the First Deep Space Maneuver (km/sec)

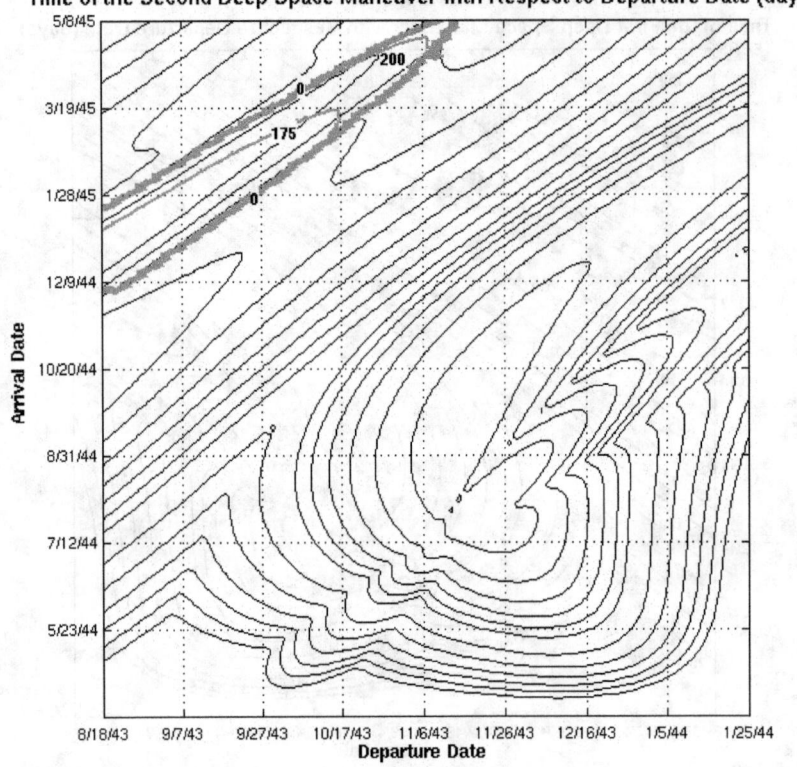

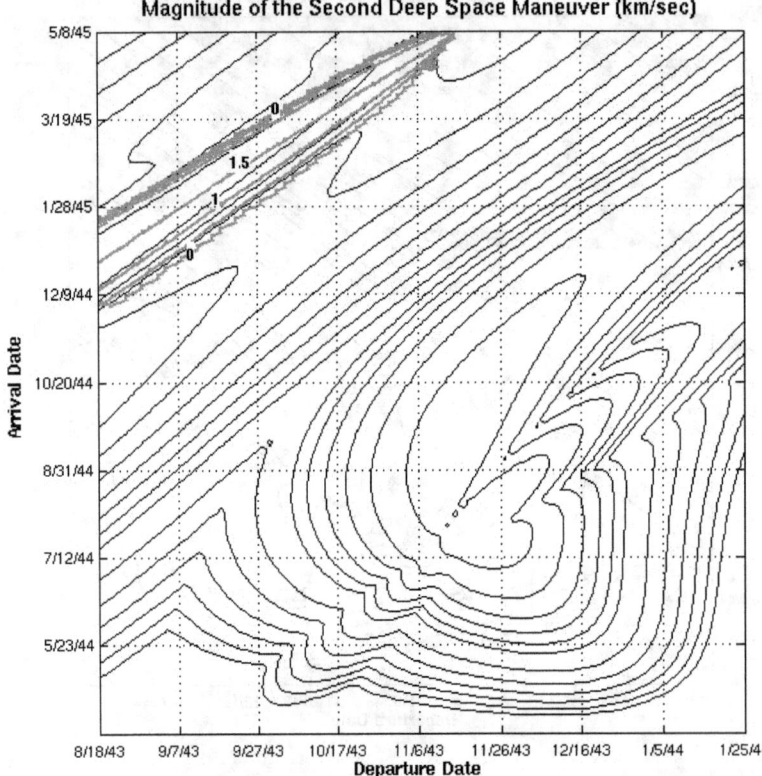

Earth to Mars—2045 Opportunity

TABLE 11.—EARTH TO MARS—2045 OPPORTUNITY—ENERGY MINIMA

Mission type	Earth departure date (m/d/yr)	Mars arrival date (m/d/yr)	C_3 (km²/sec²)	Right ascension (deg)	Declination (deg)	Mars arrival excess speed (km/s)
Type 1	1/8/46	7/27/46	**9.061**	178.80	−22.00	5.512
Type 2	1/22/46	12/18/46	**8.587**	227.60	11.17	5.119
Type 1	2/11/46	9/17/46	20.71	161.50	−14.4	**3.767**
Type 2	12/3/45	9/21/46	10.84	209.60	15.49	**3.256**

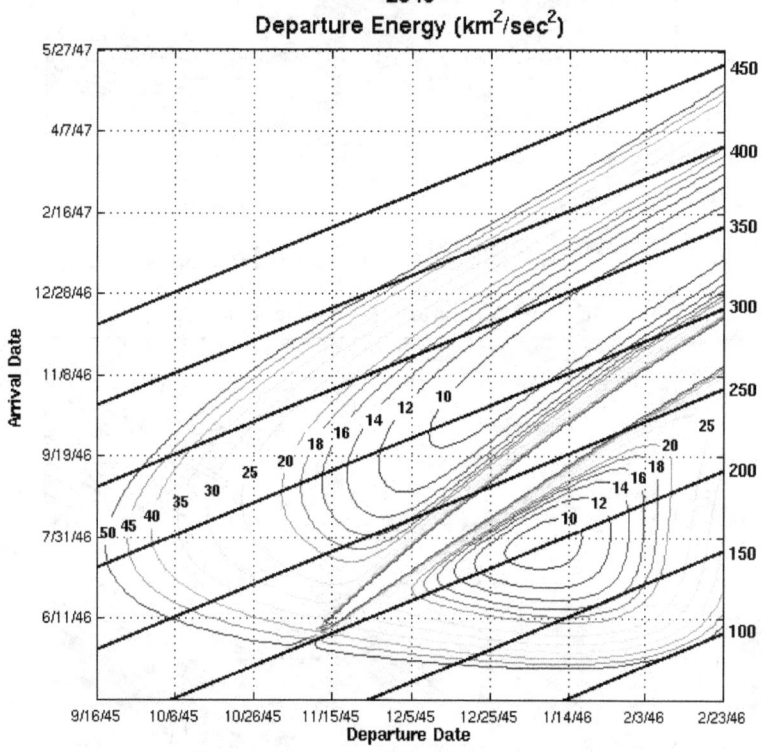

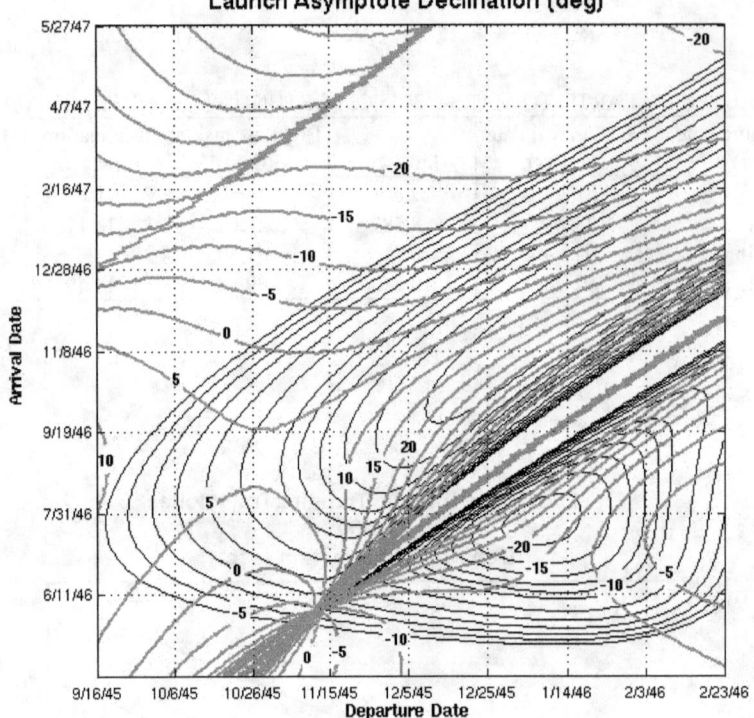

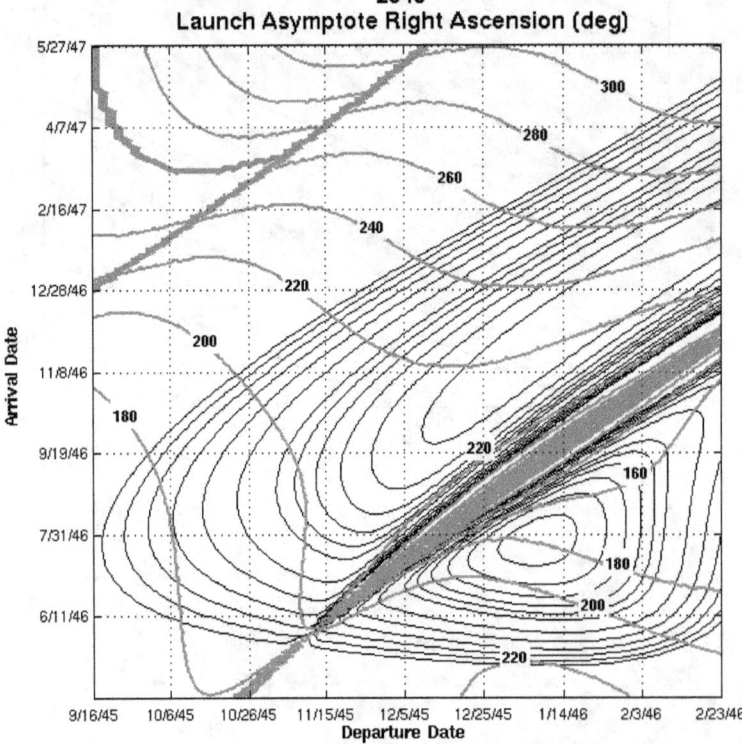

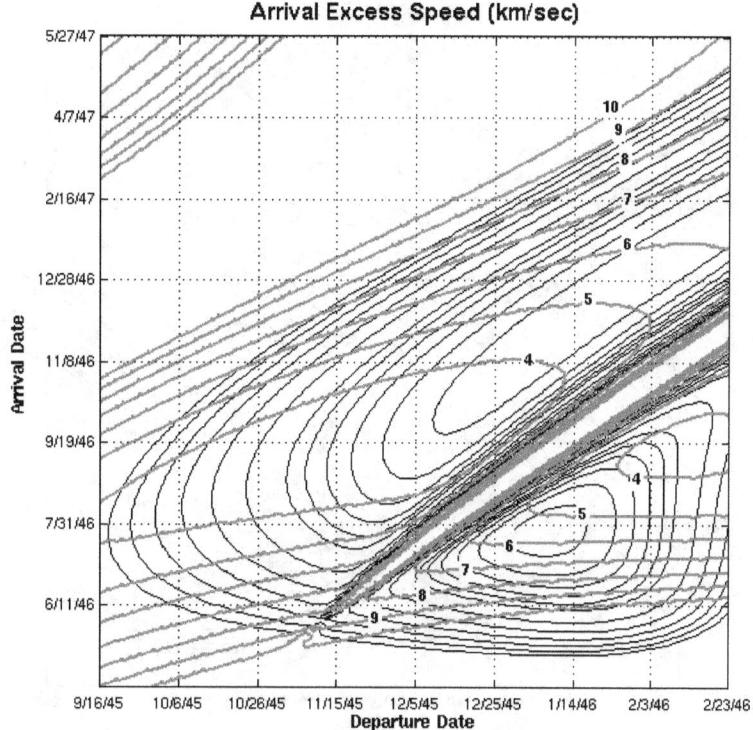

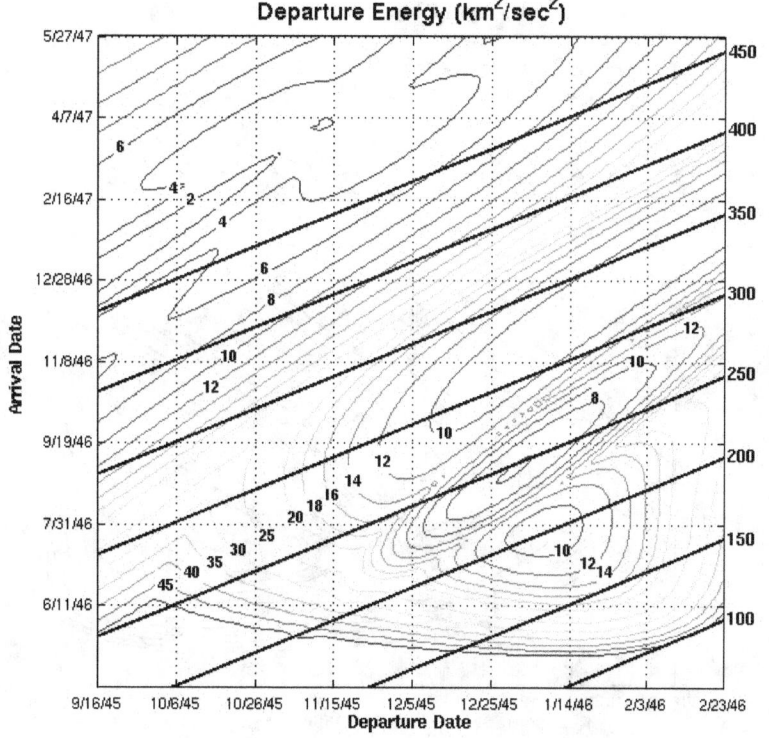

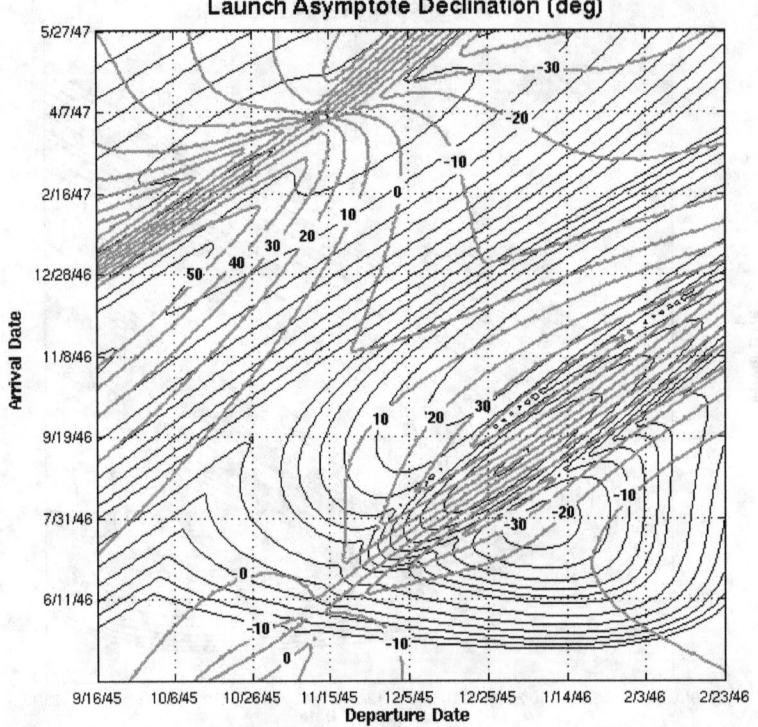

Earth-Mars Trajectories with Mid-Course Corrections
2045
Launch Asymptote Declination (deg)

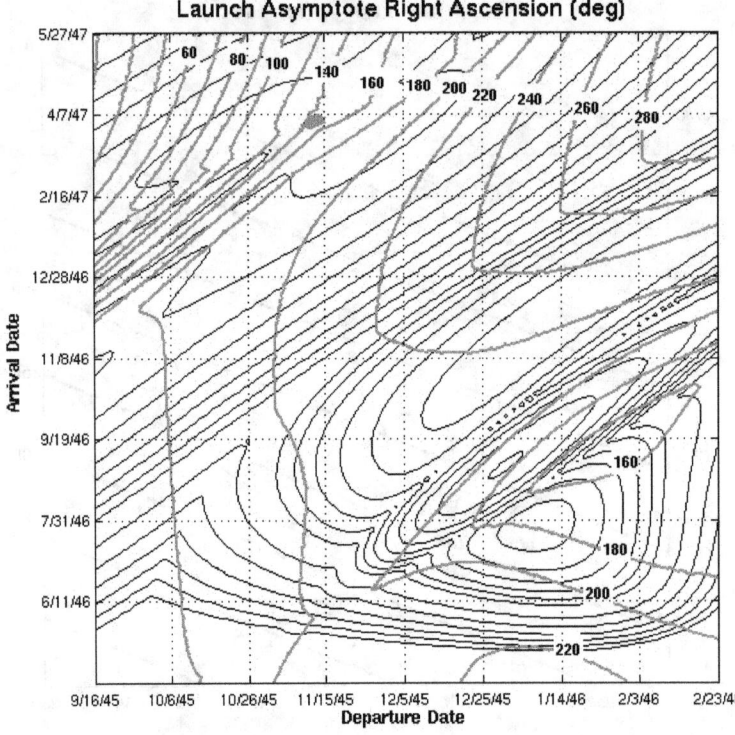

Earth-Mars Trajectories with Mid-Course Corrections
2045
Launch Asymptote Right Ascension (deg)

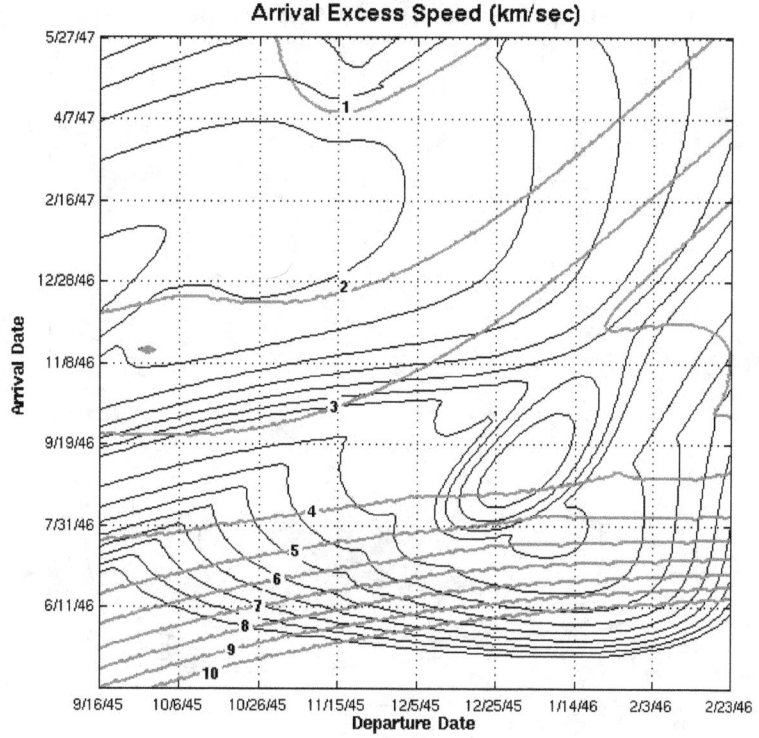

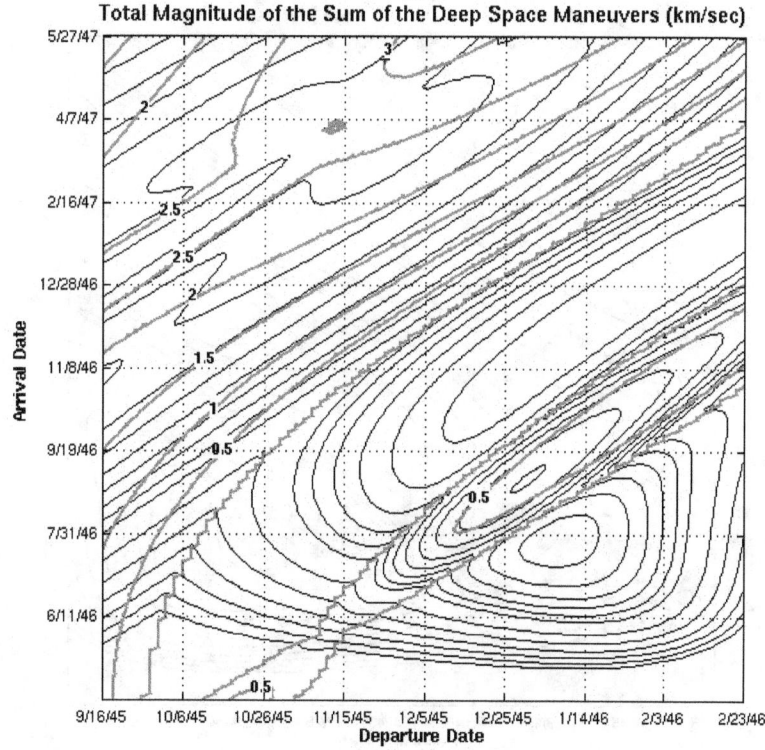

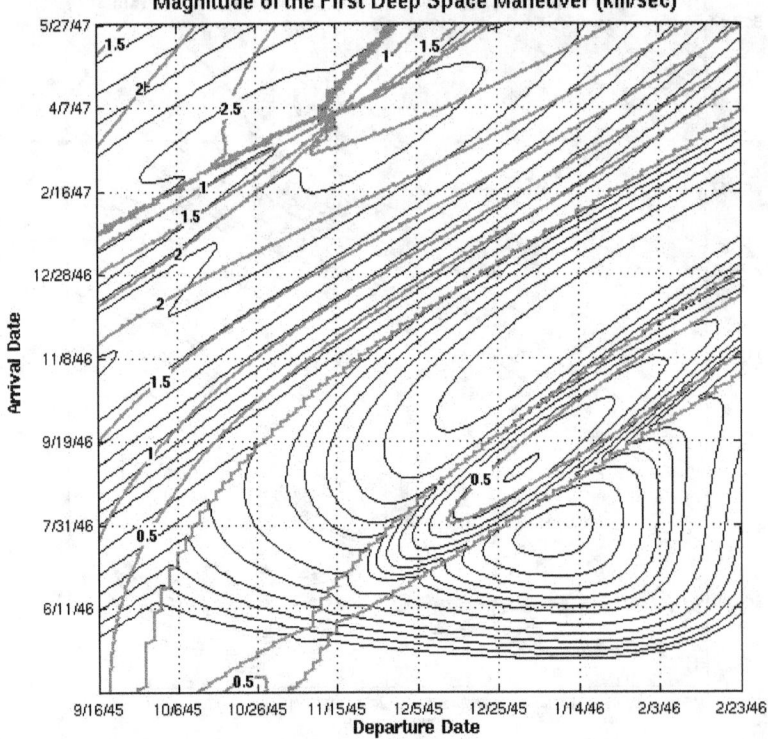

Earth-Mars Trajectories with Mid-Course Corrections
2045
Time of the First Deep Space Maneuver with Respect to Departure Date (days)

Earth-Mars Trajectories with Mid-Course Corrections
2045
Magnitude of the First Deep Space Maneuver (km/sec)

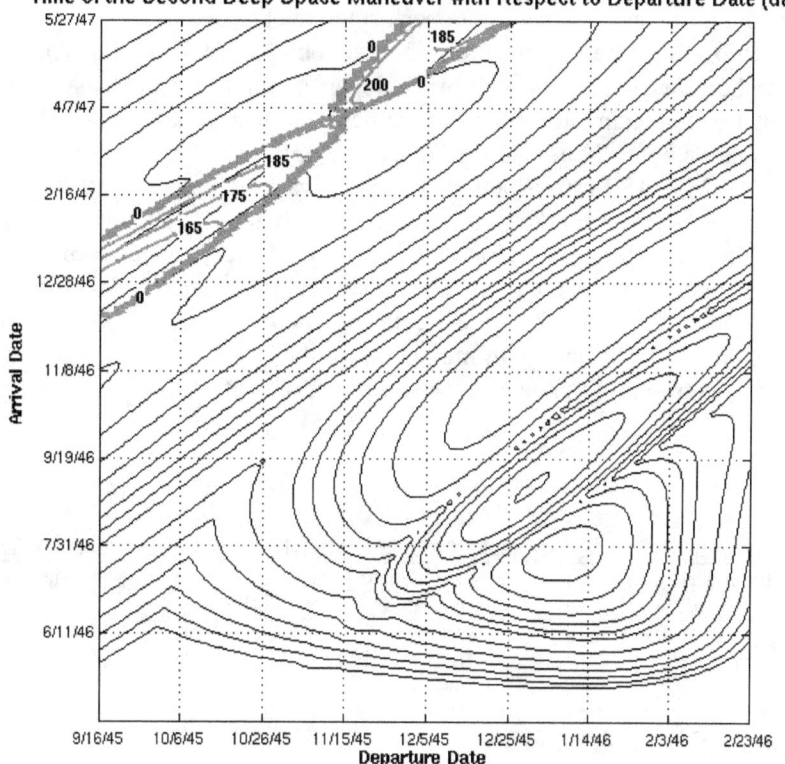

Earth-Mars Trajectories with Mid-Course Corrections
2045
Time of the Second Deep Space Maneuver with Respect to Departure Date (days)

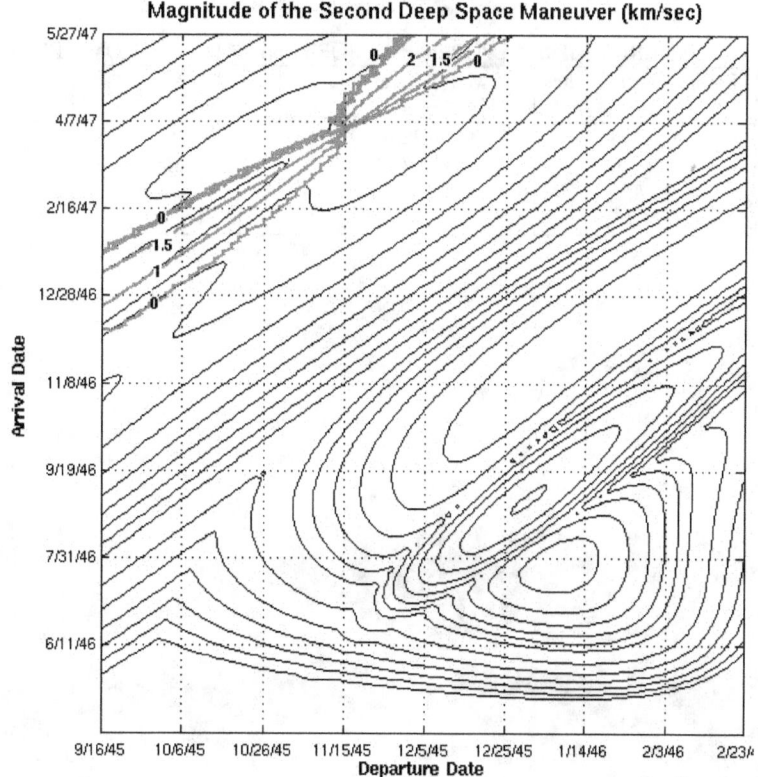

Earth-Mars Trajectories with Mid-Course Corrections
2045
Magnitude of the Second Deep Space Maneuver (km/sec)

Appendix A.—Verification of Midas Results

The MIDAS generated trajectory data with no performed deep space maneuver was validated by making a comparison between the contour plots generated by MIDAS and those published in References 3 and 6 for the year 2005. Contour plots of departure energy, launch asymptote declination, launch asymptote right ascension, and arrival excess velocity compared favorably to those in Reference 6. Reference 3 did not contain a contour plot for the launch asymptote right ascension.

The data generated by MIDAS for the 2005 Opportunity was compared to the 2005 optimum energy data provided by Reference 6. Given that the MIDAS values only have accuracy out to one hundredth of a unit, and the values provided by Reference 6 have an accuracy of one thousandth of a unit, there are potential rounding differences between the two values. The departure dates and arrival dates fall within one or two days of each other. However, because the discrepancies are small, the values are effectively equivalent. Table 12 summarizes the energy minima data for the 2005 opportunity generated by MIDAS. Table 13 summarizes the energy minima data for the 2005 opportunity provided by Reference 6.

TABLE 12.—ENERGY MINIMA FOR 2005 OPPORTUNITY CALCULATED BY MIDAS

Mission type	Earth departure date (m/d/yr)	Mars arrival date (m/d/yr)	C_3 (km^2/sec^2)	Mars arrival excess speed (km/s)
Type 1	8/10/05	2/20/06	**15.89**	3.219
Type 2	9/1/05	10/4/06	**15.45**	3.459
Type 1	9/7/05	4/19/06	25.45	**2.361**
Type 2	6/21/05	4/3/06	31.44	**2.466**

TABLE 13.—ENERGY MINIMA FOR 2005 OPPORTUNITY DATA FROM REFERENCE 6

Mission type	Earth departure date (m/d/yr)	Mars arrival date (m/d/yr)	C_3 (km^2/sec^2)	Mars arrival excess speed (km/s)
Type 1	8/10/05	2/22/06	**15.883**	-------
Type 2	9/2/05	10/8/06	**15.445**	-------
Type 1	9/8/05	4/20/06	-------	2.3602
Type 2	6/21/05	4/4/06	-------	2.4668

Verification Contour Plots for 2005 Opportunity

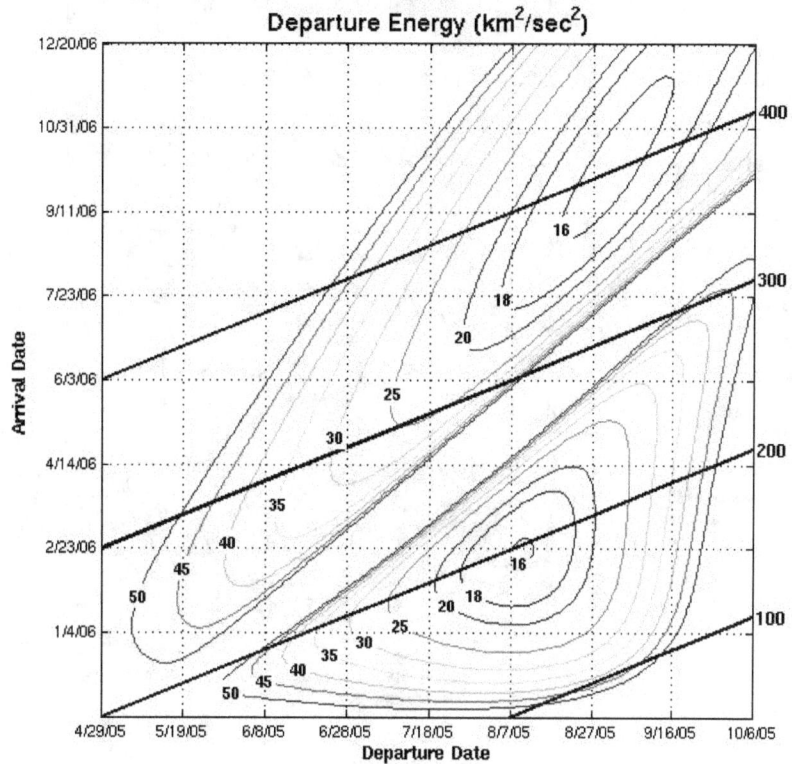

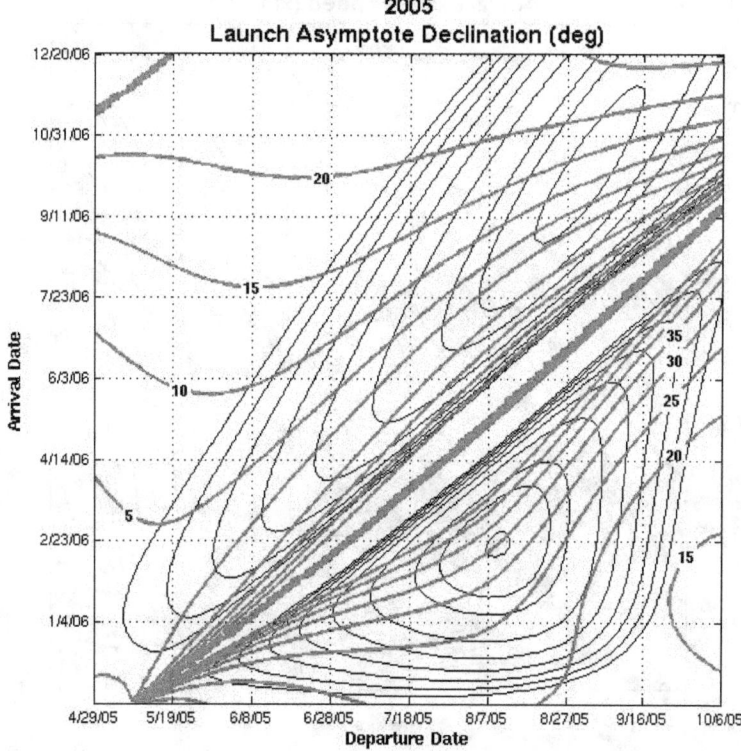

NASA/TM—2010-216764

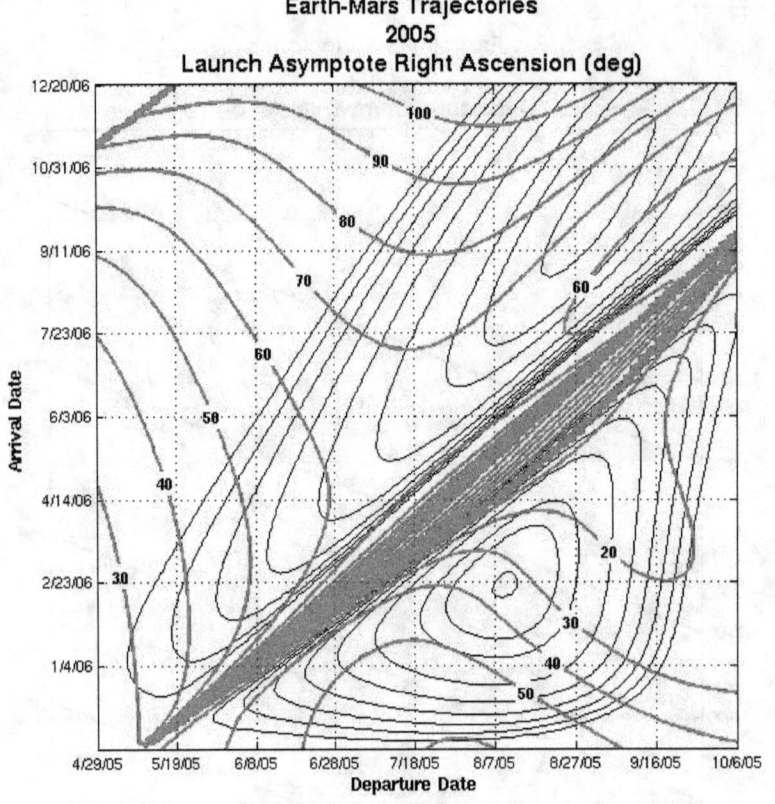

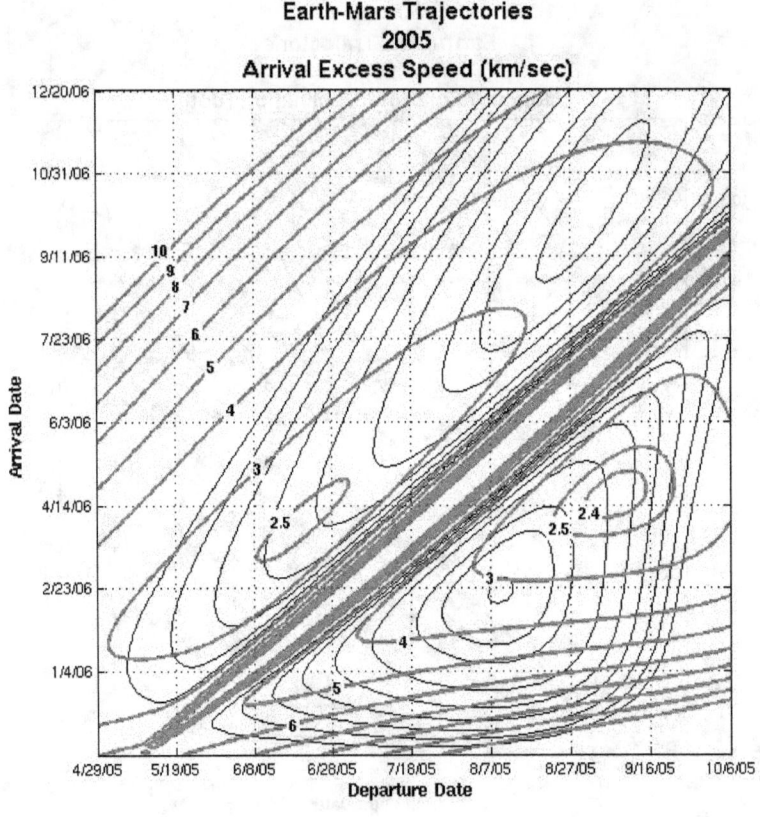

References

1. Sauer, Carl: *Preliminary Draft of Users Guide to MIDAS,* March 1991.
2. Bate, R.R.; Mueller, D.D.; White, J.E.: *Fundamentals of Astrodynamics*, Dover Publications, Inc., New York, 1971.
3. George, L.E. and Kos, L.D.: *Interplanetary Mission Design Handbook: Earth to Mars Mission Opportunities and Mars to Earth Return Opportunities 2009-2024,* NASA/TM-1998-208533, July 1998.
4. Clarke, V.C.: *A Summary of the Characteristics of Ballistic Interplanetary Trajectories, 1962-1977*, JPL Publication 32-209, January 1962.
5. Jordan, J.F.: *The Application of Lambert's Theorem to the Solution of Interplanetary Transfer Problems*, JPL Publication 32-521, February 1964.
6. Segeyevsky, A.B.; Snyder, G.C.; Cuniff, R.A.: *Interplanetary Mission Design Handbook, Volume 1, Part 2. Earth to Mars Ballistic Mission Opportunities, 1990-2005,* JPL Publication 82-43, September 1983.
7. Navagh, J.: *Optimizing Interplanetary Trajectories With Deep Space Maneuvers*, M.S. Thesis, The George Washington University, Washington D.C., Langley Research Center, Contractor Report 4546, 1993.